VALUED ENVIRONMENTS

List of Contributors

Jacquelin Burgess
Lecturer in Geography, University College London
Stephen Daniels
Lecturer in Geography, University of Nottingham
John Gold
Senior Lecturer in Geography, Oxford Polytechnic
Brian Goodey
Reader in Urban Design, Oxford Polytechnic
Derek Hall
Senior Lecturer in Geography, Sunderland Polytechnic
Susan-Ann Lee
Senior Lecturer, Architectural Psychology Research Unit,
Kingston Polytechnic
David Lowenthal
Professor of Geography, University College London
Katherine Oliver
Graduate Student, University College London
John Punter
Lecturer in Land Management, University of Reading
Marion Shoard
Formerly Research Officer for the Council for Preservation of Rural England
and author of *The Theft of the Countryside*
David Uzzell
Lecturer in Geography, Surrey University

VALUED ENVIRONMENTS

EDITED BY JOHN R GOLD
& JACQUELIN BURGESS

London
GEORGE ALLEN & UNWIN
Boston   Sydney

George Allen & Unwin (Publishers) Ltd,
40 Museum Street, London WC1A 1LU, UK

George Allen & Unwin (Publishers) Ltd,
Park Lane, Hemel Hempstead, Herts HP2 4TE, UK

Allen & Unwin Inc.,
9 Winchester Terrace, Winchester, Mass 01890, USA

George Allen & Unwin Australia Pty Ltd,
8 Napier Street, North Sydney, NSW 2060, Australia

First published in 1982

British Library Cataloguing in Publication Data

Valued environments: essays on the place and landscape.
1. Landscape   2. Nature (Aesthetics)
I. Gold, John R.   II. Burgess, Jacquelin A.
719'.01   BH301.L3
ISBN 0–04–710001–X

Library of Congress Cataloging in Publication Data

Main entry under title:
Valued environments.
Includes bibliographical references and index.
1. Landscape assessment – Great Britain – Addresses, essays, lectures.
2. Landscape protection – Great Britain – Addresses, essays, lectures.
3. Environmental policy – Great Britain – Addresses, essays, lectures.
I. Gold, John Robert.   II. Burgess, Jacquelin A.
GF91.G7V34   1982   333.7'2   81–20519
ISBN 0–04–710001–X   AACR2

Set in 10 on 12 point Bembo by Bedford Typesetters Ltd
and printed in Great Britain
by Mackays of Chatham

Preface

The idea for this book stemmed from the symposium on 'Valued Environments' that was held at the 1978 Annual Meeting of the Institute of British Geographers at Hull University. The symposium brought together a wide range of papers and attracted a large audience. The lively discussion that ensued and the general interest that was aroused convinced us, as the organisers of the symposium, that the conference papers could usefully form the basis for a collection of essays. Having said that, it should be stressed that this book is decidedly not the 'proceedings' of that symposium. During the intervening period, the essays have been extensively revised and extended from the papers that were their precursors, and several new contributions have been added to the collection. The result is a volume that has developed and matured in the process of preparation, but one which we feel retains the freshness and vitality of the original event.

The essays included here explore many different themes and tackle various types of valued environments – urban and rural, small-scale and regional, exalted and unsung – but all are bound together by their focus upon the meaning of place and landscape in the lives of individuals and groups. The essays themselves are by British authors and deal primarily with valued environments located in Great Britain, yet it is our firm belief that their findings, the ideas that they put forward, and the issues upon which they touch will interest all who are engaged in the study or design of places, landscapes and environments.

An undertaking of this sort relies upon the goodwill and co-operation of a large number of people, most of whom receive scant reward for their labours. Certainly we consider ourselves fortunate in the quality of advice, assistance, and encouragement that we have received. In particular, we would like to thank Denis Cosgrove, Alan Jenkins and Hugh Prince for their helpful comments about the essays and the book in general. None of them, of course, should be held responsible for the outcome. Our thanks go to Jay Appleton and Nigel Haigh who acted as chairmen at the original symposium. We would also like to thank Annabel Swindells and Liz Cooke for the prompt and efficient way in which they typed the final manuscript; Alick Newman for redrawing the maps, diagrams and illustrations; and Chris Cromarty for the Ideal Home in the cover photograph. Finally, as always, we wish to thank our respective spouses, Maggie and Mike, for their support and tolerance.

JOHN GOLD and JACQUELIN BURGESS
London
June 1981

Contents

List of contributors   *page* iv

Preface   vii

1 *On the significance of valued environments*   1
   JACQUELIN BURGESS and JOHN GOLD

   Concepts and contents   4
   Notes   8

2 *Values in place: interpretations and implications from Bedford*   10
   BRIAN GOODEY

   Bedford experiences   12
   Bedford by trail   16
   Place experiences   20
   Provoking place experiences   24
   Implications   31
   Acknowledgements   32
   Notes   32

3 *Filming the Fens: a visual interpretation of regional character*   35
   JACQUELIN BURGESS

   Finding the story   35
   A man–made landscape   40
   Flatness   41
   A fenland fixation   42
   Water   44
   The Fen Tiger?   47
   Isolation   50
   The sense of place   51
   Notes   53

4 *The lure of the moors*   55
   MARION SHOARD

   Primary conditions for wilderness   59
   Secondary conditions   62
   Why preserve wilderness?   63
   The shaping of public policy   65
   Notes   72

5   *Revisiting valued landscapes*   74
    DAVID LOWENTHAL

    The virtues of bygone scenes   74
    Revisiting past landscapes through science fiction   79
    Rediscoveries and reconstructions   86
    Defects and drawbacks   91
    Notes   94

6   *Landscape aesthetics: a synthesis and critique*   100
    JOHN V. PUNTER

    Landscape aesthetics: a broader field of enquiry   100
    Research paradigms in landscape aesthetics   102
    Areas of common effort and consensus   109
    Philosophy of aesthetics: alternative positions   111
    A materialist perspective: first steps in re-orientation   112
    Conclusion   117
    Notes   118

7   *Humphry Repton and the morality of landscape*   124
    STEPHEN DANIELS

    Introduction   124
    Landscape and morality   124
    Humphry Repton   126
    Sheringham   131
    Hare Street   138
    Conclusion   140
    Acknowledgements   141
    Notes   141

8   *Places, conservation and the care of streets in Hartlepool*   145
    KATHERINE A. OLIVER

    The street survey: the motifs of conservation   146
    The process and agents of change   151
    The ambiguity in streets   154
    Acknowledgements   159
    Notes   159

CONTENTS xi

9 *The value of the local area* 161
SUSAN–ANN LEE

Introduction 161
The role and value of the local area to the individual 161
The role and value of the local area in social life 163
The role and value of the local area in political life 165
Conclusion 168
Notes 169

10 *Valued environments and the planning process: community consciousness and urban structure* 172
DEREK R. HALL

The subjective reality of valued environments 172
Representation of valued environments in the planning process 179
Neighbourhood councils and valued environments 184
Notes 187

11 *Environmental pluralism and participation: a co-orientational perspective* 189
DAVID L. UZZELL

Co-orientation and a social construction of reality 190
The co-orientation model 191
Friary Ward: an introduction 193
Environmental and political knowledge 195
A case study: a residents' parking scheme for Friary Ward 197
Conclusion 201
Notes 202

Index 204

1 *On the significance of valued environments*

JACQUELIN BURGESS and JOHN GOLD

> Sometimes
>
> on fogless days by the Pacific
> there is a cold hard light without break
>
> that reveals merely what is – no more
> and no less. That limiting candour
>
> and accuracy of the beaches
> is part of the ultimate richness.[1]

The 'ultimate richness' is life as captured in the varying intensity and wide diversity of human environmental experience: pleasure gardens and city streets, rainsoaked mountains and hidden valleys, Blackpool on a public holiday and downtown Kansas City on a Saturday afternoon. We all search for environments that satisfy our basic needs: shelters in which to nurture the young and in which to die; places which afford us pleasure and mental stimulation; environments that supply an indication of our past and of what the future might hold. In seeking to satisfy these needs people will be attracted towards some environments and repelled by others, but in the process of finding their own special places they may find themselves. Biographies will become intimately connected with environments for, regardless of circumstance and position, individuals emerge to hold and create their own landscapes.[2]

In such intimate verbs as 'create' and 'hold' lies the true significance of valued environments. To *create* is to give of oneself and to endow the world which we inhabit with personal meaning. To *hold* is to cherish and offer security and protection. This immediately suggests those most cherished buildings and landscapes that are part of the national heritage and which enjoy widespread protection, but what of the other environments that supply the backcloth for everyday life: fields, woodlands and parks; village streets, suburban closes and city thoroughfares; cinemas, offices and shopping centres? Who creates them? Who cares for them and how are they held?

These are difficult questions to answer, not least because relatively little effort has been invested in trying to answer them. This neglect partly results

from the undeniable preoccupation of Western society with the environmental tastes of 'high culture',[3] but just as significant is the fact that quotidian
environments are taken for granted by policy makers, academics and inhabitants alike. The strength of attachment to ordinary places and landscapes
frequently only emerges when they are threatened by change. Gosling,[4] for
example, when writing about the redevelopment of his adopted neighbourhood of St Annes in Nottingham states that:

> Change came like a torrent sweeping all before it: houses, streets,
> chapels, shops, pubs, the whole old life. Not only the face of the
> neighbourhood changed but the body and spirit of the people: the
> constitution, grit and social composition. A history was wiped away.

Although academics are accustomed to the argument and the polemic is
one routinely encountered at public inquiries into planning proposals,
familiarity does not detract from the force of the complaint. Many people feel
deep affection for the places in which they live. Large-scale environmental
change, not instigated by the inhabitants themselves and outside their
control, often seems disturbing, disruptive and divisive. Increased demands
for active involvement in environmental decision-making, the burgeoning
growth of pressure groups, and widespread suspicion of policy-making
bodies together reflect a belief that political decision-makers cannot and will
not do enough to protect our environmental interests.[5] In particular, criticism
is directed against the assumption that policy makers acting alone can
determine and safeguard the 'public interest'. Indeed, in many instances, the
true nature of the 'public interest' remains obscure, a fanciful conception
occasionally invoked rather than a goal to be actively sought. Clearly the
public interest is not served by the activities of individuals whose pursuit of
quick profits leads to the irrevocable destruction of important historic
buildings or wildlife sanctuaries, but equally it is not necessarily promoted by
the imposition of schemes that ignore the needs of local communities in order
to serve some notional concept of the common good of society at large.

Who then is to say what constitutes the 'public interest' in terms of
environmental quality and protection? The traditional arbiters – poets,
painters, novelists, wealthy travellers, politicians and landowners – have
now been supplemented by newer elites such as the mass media and pressure
groups who similarly represent sectional interests. Their views and counsels
undoubtedly carry much weight, but may well be an inadequate guide to
understanding the nature of environmental experience and values. Elite and
public preferences for places and landscapes may coincide. There may be a
time lag during which elite taste diffuses into a public consensus, but there are
also many instances in which preferences conflict.[6] The position of the
conventional 'authorities' in environmental matters is being challenged by a
large and increasingly well articulated body of public opinion emanating

from people who care deeply about the future of their environments, even if they have no pretensions about the depth of their knowledge.

Given the movement for greater public participation in planning, which now has statutory force in some countries,[7] the planner can respond in various ways. One way is to retreat behind a smokescreen of 'professionalism', as was highlighted by the response of Sir Desmond Heap[8] to a debate on British planning policy:

> What point could come from a professional man discussing his professional expertise with a layman I really do not know because it is perfectly clear (is it not) that the layman, being uninformed and untaught in the mystique of the particular profession in question, would simply not know what on earth he was talking about.

Even more strident positions have been adopted by others in defence of the professional's freedom of action. Allison,[9] for example, argued that: 'We all know what people want from their environment. They want privacy, a good view, a short journey to work, an appreciating property, convenient retail outlets, adequate recreational facilities and a beautiful landscape.' The short answer to such cynicism is that we do *not* know what people want. We are far from being able to say what constitutes a good view, a beautiful landscape or adequate facilities, nor have we found satisfactory means by which to articulate the nature and significance of the intimate attachments to place and landscape that will be encountered at many stages in this book.

Both academics and planners have attempted to fill this void by recourse to the methods of social investigation: private feelings become public and accessible through the use of questionnaires, scaling tests, landscape inventories, photographic surrogates and simulations.[10] This is a worthwhile and valid endeavour but it is doubtful that existing techniques provide a necessary and sufficient understanding of the environmental experiences and values of others. Social survey techniques may be criticised for their basic assumptions, their lack of theoretical underpinning,[11] and their encouragement of a false consciousness in which the independent observer makes detached judgements about the feelings and values of inhabitants or the general public. Above all, it has to be recognised that many people find it hard to articulate their own feelings about place and landscape. Describing what an environment feels like by means of simple test procedures can reduce the quality of the experience to platitudes and clichés; as Huxley[12] put it: 'touch the pure lyrics of experience and they turn into the verbal equivalents of tripe and hogwash'.

Many writers[13] have suggested that the sense of detachment and superficiality produced by some applications of social science techniques could be avoided by returning to the locality and building up a detailed and sympathetic understanding of the values and desires of local inhabitants. This strategy has at times yielded extremely useful insights and, as several contributors to this book indicate, is an important avenue for research. However, there are dangers inherent in what may well become a 'cult of localism'. Lynch's exhortation[14] that academics and environmental managers should live in the locality and talk at length with local people to uncover their deep values and beliefs is admirable, but is it likely to be taken seriously by those charged with the immediacies of executing practical planning policy? Time is of the essence: the bulldozers wait impatiently to dig up hillsides and flatten buildings. In their wake may be a place that has lost irretrievably the essential qualities that once made it valued. Equally, justifiable concern with the locality and the environmental wishes of its inhabitants must not be allowed to degenerate into preoccupation with the *status quo* and preservation in opposition to all new development. Present environments are themselves the result of continual, incremental change. Preservation seeks to fossilise environments and it denies creativity: it is life-denying rather than life-enhancing. Whether done by legal ordinance or financial assistance, preservation can only retain the fabric[15] and, if that fabric cannot accommodate present-day needs, then the environment invites its own destruction.[16]

The task therefore is to find perspectives and approaches that allow us to identify the true nature and significance of places and landscapes that are valued, without losing sight of either the wider context of society as a whole or the pressing need to address our inquiries, at least in the longer term, to practical as well as academic ends. This spirit of investigation underpins the essays contained in this book. Before describing them in greater detail, it is first necessary to introduce the central concept that binds them together.

Concepts and contents

In his introduction to a distinguished set of essays on the interpretation of ordinary landscapes, Meinig[17] noted that the key notion that linked them – landscape – was at once attractive, important and ambiguous. Much the same can be said of the idea of a 'valued environment'. The term is attractive in that it provides a point of convergence for the study of many diverse environments. As the contents of this book clearly indicate, it can embrace with equal facility the favoured landscapes of the elite and the unspectacular environments in which most of us live, the enchanted places of childhood and the harsher realities of adult life, the sweeping moorlands of upland Britain and microcosms of inner urban neighbourhoods, the contemporary and the historic, the commonplace and the unique. This broad coverage is itself extremely helpful but it should be stressed that 'valued environments' is more than a convenient umbrella term. It is an important concept in that it refocuses attention upon the close and enriching affective bond between people and the environments that they create, inhabit, manipulate, conserve, visit or, even, imagine. In particular, the notion of valued environments emphasises the benefits that are to be derived from association with, attachment to, and love of, certain places and landscapes: a perspective that has helped to balance the 'bias in favour of mobility' that Lenz-Romeiss[18] has discerned in much contemporary social science theory. However, the ambiguity inherent in the concept needs to be clarified before the authors of this book begin to capitalise upon its enigmatic quality.

'Environment' itself contains an important ambiguity of meaning. Although often used to denote a specific portion of space, it has a variety of associated meanings as an 'ambience' and as 'the conditions under which any person or thing lives or is developed'.[19] Hence environment also connotes social and developmental properties which in turn relate to such questions as the emergence of personal identity and socialisation. Moreover, the word 'environment' is not tied to any specific geographical scale. Rooms, dwellings, localities, neighbourhoods, towns and regions can all be regarded as being encompassed by the term. The adjective 'valued' merits rather fuller discussion. What is it that makes an environment valued? Are there general features and processes that endow an environment with this quality or is it merely the product of singularity? Is value to be associated principally with the gradual development of intimate attachment or is it more the result of sudden revelation? Can the nature of a valued environment ever be fully expressed or must it always be ultimately unknowable and approachable only through painstaking attempts to build up a sympathetic understanding? Resolution of these questions depends upon appreciating the complexity of the word 'value'.

In the first place, there are two contrasting theoretical approaches to the study of value.[20] One view sees value as an absolute quality, resting on the assumption that worth is intrinsic to the entity itself independently of the context in which it is found. The other view sees value as a relative quality, assigned to an entity on the basis of comparative assessment against other entities and dependent upon the context in which it is found. The ancient fable of King Midas illustrates that this is not a new distinction, and also perhaps indicates that the second position is closer to human experience and behaviour than the former.[21] Certainly the notion of 'relative value', with its requisite measure of comparative assessment, is the dominant view taken in the ensuing essays.

Secondly, and despite the recent advertising campaign of a major British newspaper under the slogan 'Times Change, Values Don't', value *is* a dynamic concept. The literature of architectural history and landscape gardening is replete with many instances of oscillations in taste for different types of building, artefact and landscape. Another good example is supplied by our continual reassessment of the value of buildings of different historic eras. Victorian buildings, once condemned as ugly and full of 'bogus' pseudo-Gothic ornamentation, have now been reinterpreted as having value in their own right; an appreciation that goes well beyond the middle-class *cognoscenti* who normally form the backbone of Victorian Societies. Indeed the recent appearance of a Thirties Society dedicated to the preservation of Art Deco factory buildings, cinemas and artefacts demonstrates that the process is continuing apace.

Finally, it is essential to know exactly by whom the environment is supposed to be valued. As has already been seen, there may be a considerable lacuna between elite and public environmental values, although it is also possible that these values may coincide or converge if viewed over a longer period of time. The relationship between values at the individual, group and cultural levels is complex and multi-faceted. Certain values overlap and provide a degree of consensus between disparate groups of people, whilst others are unique, specific and idiosyncratic. We may assume, for example, broad agreement about the significance of Stonehenge as part of Britain's national heritage, but it is far more difficult to identify those for whom the small town high street or coppice represents a 'valued environment'.

As befits the concept of valued environments, the contents of this book are richly diverse in theoretical, empirical and methodological perspectives. Individual experiences with places, insights into other people's affections for environments from conversations and interviews, interpretations of valued landscapes in imaginative literature, data from questionnaire surveys, information from historical sources, academic debate – all have a part to play in illuminating the meaning of place and landscape.

The first three essays are linked through the authors' emphasis on experience, either reflecting on their own experiences in coming to terms with places or through discussion of the environmental experiences of other people. The opening essay explores the use and effectiveness of town trails as a means of interpreting and communicating the value of places. Brian Goodey, a leading figure in the development of urban trails, allows himself the pleasure of a weekend in Bedford unencumbered by any academic preconceptions. He compares this freewheeling experience with the formalised routing of a town trail and suggests that many interpretative exercises are being killed through a desire to instruct rather than to provoke curiosity and enhance a person's sense of feeling for the place. This sense of feeling and affection is uppermost in the next essay. Appointed as consultant for a television documentary film, Jacquelin Burgess describes the search for a sense of place experienced by the inhabitants of the English Fenland. In identifying and illustrating three themes which communicate perceived qualities of the regional character, she develops a detailed picture of the Fens and clearly demonstrates that the meaning of the ordinary is rarely obvious – even on television. Moving from the flat land of the Fens to the moorlands of upland Britain, Marion Shoard examines the influence of elite preferences for wilderness on public policy for countryside conservation. From lengthy conversations with people who are active and influential in the conservation movement, she identifies reasons why individuals feel such deep affection for moorland and demonstrates how the sensibilities of the few influenced the designation of the British national parks in moorland areas.

The following essays by David Lowenthal, John Punter and Stephen Daniels turn our attention more closely to questions of values in landscape. First, David Lowenthal draws on the exotic literature of science fiction to illustrate the affections that people feel for landscapes redolent of the past. He demonstrates the various devices through which imaginative literature allows people to revisit past landscapes and the motives that lie behind this fascination. The author concludes that all efforts to recapture valued landscapes, whether fictional or visionary, are doomed to failure because they are contingent on our own views: past environments and places conform to our present conceptions of them. Expectations and assessments of beauty in contemporary environments provide the substance for John Punter's review of the field of the landscape aesthetics. After categorising the available perspectives into basic research paradigms, he critically examines their underlying philosophical orientations. He concludes by advocating a materialist approach which serves as a means for redefining the preoccupations of landscape aesthetic studies, as a vehicle for critical synthesis of current developments, and as a way of identifying future research questions. Stephen Daniels continues the theme of examining valued environments through landscape aesthetics, although from a rather different standpoint again. He explores the means through which landscape gardening in Georgian times gave visible expression to aesthetic and moral values. Through a discussion of the work of Humphry Repton and, in particular, his commission at Sheringham, Stephen Daniels shows the way in which values and ideals interpenetrate in the creation of landscapes.

The remaining essays are bound together by a common appreciation of the value of locality in everyday urban life. Katherine Oliver visits the industrial town of West Hartlepool and, through a detailed study of the major streets in the centre of the town, builds up a picture of the relationship between people and place. She identifies a number of discontinuities between the local authority's attitudes to conservation in the town and those of the inhabitants, and makes a plea for more personal freedom in the design and ornamentation of streets. Following this, Susan-Ann Lee suggests that attachment to local area is a significant aspect of individual relationships with places. She points to the widening gulf between planners and the planned and argues for closer integration between councillors, administrators and inhabitants. The isolation of elected members from the everyday residential experiences of their constituents is developed further in the essay by Derek Hall. He identifies valued environments in the city of Portsmouth as being those areas for which residents feel strong attachment and where community consciousness is well articulated. However, Derek Hall is less certain that the concept of valued environments is useful for planners since it contains so many ambiguities in meaning. His challenge is answered in the final essay where David Uzzell shows how the concept may be put into action by means of the technique of co-orientation – a method of drawing out shared meanings among different groups of people. After a critique of existing methods used to discern public feelings and attitudes, he outlines the possible advantages of co-orientation and illustrates them by means of a case study.

We began with an extract from 'Flying above California' by Thom Gunn, in which he describes the intense pleasure of experiencing 'merely what is' in landscapes. In these essays, our authors have drawn primarily on British examples to articulate something of the unpretentious and everyday qualities that people value in places and landscapes. We hope that our readers, perhaps residing in different environments and contrasting cultural milieux, will be stimulated by these explorations. Individual experience, emotional attachment, a sense of control and a duty of care are vital aspects of living in the world.

Notes

1 Gunn, T. 1974. Flying above California. In *Worlds: Seven Modern Poets,* G. Summerfield (ed.), 85. London: Penguin.
2 Samuels, M. S. 1979. The biography of landscape. In *The Interpretation of Ordinary Landscapes,* D. W. Meinig (ed.), 51–88. Oxford: Oxford University Press.
3 Meier, R. L. 1980. Preservation: planning for the survival of things. *Futures* **12,** 135–6.
4 Gosling, R. 1980. *Personal Copy,* 195. London: Faber.
5 See *Vole,* January, February and March 1980 for a series of articles on environmental pressure groups.
   Sandbach, F. 1980. *Environment, Ideology and Policy.* Oxford: Blackwell.
6 For example, Perin, C. 1970. *With Man in Mind.* Cambridge, Mass.: MIT Press.
   Brolin, B. C. 1976. *The Failure of Modern Architecture.* London: Studio Vista.
   Appleyard, D. 1979. Introduction. In *The Conservation of European Cities,* D. Appleyard (ed.), 8–49. Cambridge, Mass.: MIT Press.
7 Fagence, M. 1977. *Citizen Participation in Planning.* Oxford: Pergamon.
8 Response to Garner, J. F. 1979. Skeffington revisited: policy forum. *Town Plann. Rev.* **50,** 425.
9 Allison, L. 1975. *Environmental Planning,* 69. London: George Allen and Unwin.
10 See Michelson, W. H. (ed.) 1976. *Behavioural Research Methods in Environmental Design.* Stroudsburg, Pa: Dowden, Hutchinson and Ross.
   Penning-Rowsell, E. C., G. H. Gullett, G. H. Searle and S. A. Witham 1977. *Public Evaluation of Landscape Quality.* Report 13. Enfield: Planning Research Group, Middlesex Polytechnic.
11 Appleton, J. 1975. Landscape evaluation: the theoretical vacuum. *Trans Inst. Br. Geogs* **66,** 120–3.
12 Huxley, A. 1955. *The Genius and the Goddess,* 36. New York: Harper and Row.
13 For example, Buttimer, A. and D. Seaman (eds) 1980. *The Human Experience of Space and Place.* London: Croom Helm.
   Mikellides, B. (ed.) 1980. *Architecture for People.* London: Studio Vista.

Jencks, C. 1977. *The Language of Post-Modern Architecture*. London: Academy Editions.

Relph, E. 1976. *Place and Placelessness*. London: Pion.

14 Lynch, K. 1976. *Managing the Sense of a Region*. Cambridge, Mass.: MIT Press.

15 Tunnard, C. 1978. *A World with a View: an Inquiry into the Nature of Scenic Values*, 141. New Haven, Conn.: Yale University Press.

16 Lynch, K. 1972. *What Time is this Place?* Cambridge, Mass.: MIT Press.

17 Meinig, D. W. 1979. Introduction. In *The Interpretation of Ordinary Landscapes*, D. W. Meinig (ed.), 1. Oxford: Oxford University Press.

18 Lenz-Romeiss, F. 1973. *The City: New Town or Home Town?* Translated by E. Küstner and J. A. Underwood. London: Pall Mall.

19 Oxford English Dictionary (compact edition) 1971. Vol. 1, 880.

20 Facione, P. A., D. Scherer and T. Attig 1978. *Values and Society: An Introduction to Ethics and Social Psychology*. Englewood Cliffs, NJ: Prentice-Hall.

21 Cooper, J. B. and J. L. McGaugh 1966. Attitude and related concepts. In *Attitudes* M. Jahoda and N. Warren (eds), 30. London: Penguin.

2 Values in place: interpretations and implications from Bedford

BRIAN GOODEY

Environmental interpretation, as a discrete and near-professional activity, has emerged as an adjunct to recreation and tourist management, environmental design and education in the past few years. In earlier papers,[1] I have outlined some of the issues raised and opportunities provided by the application of predominantly rural-based techniques to urban conservation and to the process of urban design. With the recent publication of the Civic Trust's manual on urban interpretation,[2] the efforts made towards achieving recognition for the activity might be seen as having been rewarded.

Having been involved in the promotion of 'the town trail movement' by listing the many trails that emerged in Britain and developing guidelines for trail preparation and use,[3] I became concerned at the pattern-book form of many published trails. Production of these guide-leaflets seemed to be dominated by local historians and architects, and although one hoped that each trail would broaden the user's experience of place, there now appears to be every danger that the cultured stereotypes of places as mere background to a few impressive buildings and historic associations will be reinforced and persist.

The dominant 'heritage' interest in our society diverts attention from contemporary places. Few writers or television presenters have the ability to evoke the spirit of our towns as they are, rather than as they should be. Ian Nairn, in his emotional and idiosyncratic excursions, provides one exception to this generalisation; Jan Morris, Colin Ward, James Cameron, even John Betjeman, have hinted at the larger world beyond the treasured structures of a place. None of these authors achieves recognition in the Civic Trust's new manual, rich though it may be in castings for metal plaques and brochure printing.

In the essay that follows, I describe an exercise in mental – and at times physical – jogging undertaken in the autumn of 1978. Concerned lest my own perception of places became stylised by continued exposure to piles of 'interpretive' literature, I decided to visit a novel environment and examine it first through unguided experience, and then through the medium of an available town trail. I would forsake the usual background research and experience 'place' as a trained, but empty-handed, observer allowing the place itself to 'say the first word'.

The notes in the first two sections are taken from a two-day journal on Bedford, a town I had passed through only twice before and about which I knew only John Bunyan, bricks and an indefinite location between where I came from (Chelmsford) and where I now live (Banbury). I was conscious of taking with me an interest in market towns, in sketching and in writing, and two unread Bedford town trails. Only after the visit did I realise that the spirit of Reyner Banham[4] and of Mass Observation in the 1930s and 1940s had also been packed.[5] On arrival I decided to spend the Saturday (market day) in recording my experiences of place, and the Sunday in walking one of the trails (Fig. 2.1). The two sets of observations reported below are based on notes taken at the time and are a record of my transactions with place.

No claim is made for novelty in this approach. I needed a strategy for getting in touch with self, and with place. That this can happen to others was recently borne out by the discovery of Clare Cooper Marcus's essay on Minneapolis.[6] Returning to Minneapolis after a long absence, Ms Marcus sat, observed, and noted in two newly designed public spaces in the city. Her report is more structured to meet a casual inquiry into design and behaviour than are my own notes, but the desire to report experience of place is strongly in evidence.

Bedford experiences

(1) What goes on and/or buying a sketch pad. Hotel information? *'None, try the local paper.'* Tourist Office in Town Hall, well marked but only open 9–5 Mon.–Fri. and on the 5th floor; weekend Tourist Office in Bedford Museum (save for later). If you were 12 feet tall you'd see that the flower bed spelled out '1678 Bunyan's Pilgrim's Progress 1978'.

Arcades, arcades, The Arcade, West Arcade. 'Jerry Bol – The World's Largest One Man Band' looks far too straight and is playing his Thermos flask. Newsagent – here the town falls off very quickly. 'SORRY NO BEDFORD TIMES TODAY – WORKS DISPUTE' but there are three copies of *Corriere Della Sera* in the rack – and an Italian ice-cream seller.

For an art shop first seek the wool shop: *'Well, there's two, but, well the one in Greyfriars . . . sort of somewhere beyond the bus station . . . how can I explain it?'* She did, I managed. Signs make the street, signs and people. 'A. R. Lindley & Son, Lucky Wedding Rings Depot' in beige with a faded horseshoe, dating from the time when there were such things. Opposite, the multicolour continuity of FLASH/FLASH/C & A and next door the Co-op flashes more and offers less.

Increasingly homogeneous facades in a 1950s–1960s acceptable style.

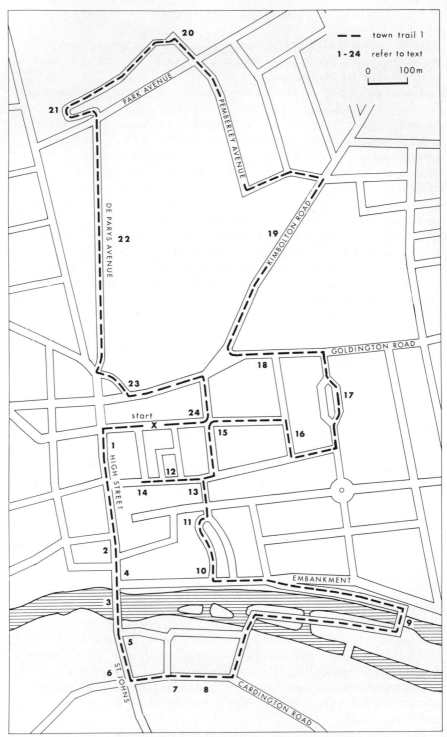

Figure 2.1 The Bedford urban trail.

'Snooker Centre' – it will never make it in those modern clothes; they have to be seedy, despite the 'Pot Black' front.

Falling off the edge again near the bus station and police tower – art shops seem not to have an obvious place, unlike junk emporia or building societies. Difficult to ignore the polite blocks of new housing here – offensive, but in-scale offensive; dead space and landscaping around the police head-quarters. Trains nearby, faces less animated, they haven't lit up for the town centre stroll yet. A *hard* surface play-area-cum-park with 'equipment' coded red for danger; *'Stand still now'* – no crossway between here and the estate opposite.

North Parade, Greyfriars – not the place for an art shop. Brick, iron balconettes and the yellow, grey and cream Festival of Britain mural:

'Drawing pins? We haven't got any at the moment but we usually do.'

(2) Bus station. Bus stations either work, or they are dead. Here at midday Saturday with the square of shops in action and the station a crossing on a thousand paths, it works; information is abrupt, signs are efficient but 'austerity'. The swath of green United Counties coaches gives away the moving map of Bedford's hinterland – ST. IVES, HITCHIN, BIGGLESWADE, BROMHAM. Do they knit together the meanest, unknown places? – from the faces they do.

Market towns are measuring devices for Middle England's changes. A proud Rasta hairdo, too many average families with polite kids and recently purchased Wrangler sweatshirts, all replicate the pace of a media world; but there remains the foundation of caps, tweed, lovat trousers, brogues and walking sticks, the straw bound grey perm and the neutral toned raincoat of those whose style stopped before it began, those for whom a day in town is ritual. Bench sitting, mulling over the pace of change.

This place is stuffed with people, all neatly distributed through tight buses-opening-and-closing views, making angles, honeypots to orderly queues which tease out the knots of blue – everyone wears blue it seems. HITCHIN is off, stolid faces set to endure the cost of the return, back seat teenagers lighting up.

Some go further: cases on wheels, the shattered look of pre- or post-holiday extraction. Brush off the seat, make a secure patch in the jungle. Jungle there is if you look beyond the people – an exercise in street furniture endurance, rails fit for cattle, posts at angles, litter bins, more posts, and overall the landscape architect's tree cover, sycamores survive the manscape and meet the buses to green and soften over all.

Somehow there is modest excitement in the rogue bus – not the standard offering but a WALLACE ARNOLD thrusting out to the coast – a long way off wherever you look for it. And the lilac or apricot of rural operators using routes, vehicles and house colours which the big boys will not touch – and making a profit on market day runs. How rituals and responses survive, how past grace survives in aged bodies making a display or infirm movement.

(3) Jerry Bol – The world's Largest One Man Band. Backed by the house sign for Peter Lord, the one man band is now in motion. 'Danny Boy' is moderate with feeling harmonica and banjo, but the dreaded 'Smurf Song' causes overload problems with a great surge of bells, banjo and drum: there seems to be real danger of lift-off like some Emu-plumed Harrier jump jet. Few dare stop to become engaged in the operation – two feet of pavement is kept clear between performer and passers by as children are dragged past the outrageous performer and parents do the classic British imitation of a moose, unsmiling, brave that they indeed have passed by. Only the kids stop, innocently amazed – 'Lavender's Blue'. How does it all work? What does it all mean? Read the labels, probe the face, it's magic on Saturday. Foreign voices respond, but *'there used to be a nice cake shop there'* elocutes a lady as she ignores what has taken its place. Now a crowd gathers, kids firmly planted, the passage blocked – but there are smiles, we cannot afford not to stop, not to be part of a communal event. Wording this nearby, I am now overlooked – an extension of the band: 'How Much is That Doggy in the Window?' wipes them out as the dog squeaker barks in the money – smiles are arcade-long. 'The Runaway Train' with real whistle builds on the audience, smiles, confusion: an event, an EVENT, sufficient to talk about, and the crowd sticks.

(4) Museum and art gallery. I trip over the museum. Entrance army of prints and drawings 'foxed'; random Victorian water colours and, round the screen, a large room (the museum it turns out) full of cases and objects: *item* 'Bedfordshire Geological Model': *item* (serried ranks of) 'Partly reconstructed pots': *items* (but one shelf apart) 'Cromwellian Mario and Breastplate found locally on a farm': *item* 'Stoneware Jar (Gent & Armitage) Bedford 1909'; *item* 'Early Radio Set (receiver) by Igranic, Bedford': *item* 'Very Early T.V. Set *c.* 1949': *item* 'Crucifixion Group made in bone at Chelmsford Gaol by French prisoners taken in 1815 at Waterloo': *item* 'Lock and key from chest of "Capt" Henry Morgan notorious pirate': *item* 'Chain carved from a single piece of wood. Location unknown'.

   Set of Sepia Views of Old Bedford (50p): The Siege of Bedford Castle (15p): Town Origins and Development (50p): Bunyan Postcards (5p): Paperweights (£1·50).

   Cecil Higgins Art Gallery – well public-gardened entrance to the ubiquitous Victorian house in grounds, overlooked by Edwardian homes of note. The garden a nice place to go – modestly graffiti-covered benches each endowed with two ladies in sensible shoes and summer frocks massaging their faces as they offer hushed descriptions of operations past, future or imagined.

Even with its extension, what can we expect from an art gallery after such a museum? Victoria rules OK? Too many attributions suspected, captured from foreign travels by local worthies; one or two surprises and either an amazing white room full of modern prints (on loan) or nothing admitted since Augustus John and Alfred Munnings. Flash Fresh – the new wing is 1976 (still looks 1956 outside) but well-entranced with stocked shop, a note on the foundation and evidence of children at work. Still people walk as if at a funeral. HENRY MOORE (b. 1898) 'The Helmet', 1950 bronze – purchased 1961 with the generous help of the Artist. BARBARA HEPWORTH (1903–75) 'Four Figures Waiting', 1968 polished bronze – purchased 1970, with a 50 per cent government grant.

Turner watercolour. List of significant events including earlier criticism of purchasing policy. Museum and gallery are chalk and cheese. An energetic Rowlandson – Fuseli, Blake, Cotman, each one a good example; the Victorians creep in – Alma-Tadema, Burne-Jones, Rossetti and company and inevitably, Augustus John; amazing bed and table by William Burges – mid-Victorian Pre-Raphaelite/Pennsylvania Dutch red/gone wild; collection of Charles and Lavinia Handley-Reed who knew their stuff and owned a national treasure nearby; Dadd's 'Madonna and Child'.

Move on to furnished rooms with the creaded, twinkling, mock-wood electric fire – but upstairs glass and porcelain of high quality, thoughtfully exhibited, Tiffany and (in Betjeman voice) *'Hurray for Charles and Lavinia Handley-Reed.'* Gather ye postcards whilst ye may – Burra's 'Harlem' which has been on my office wall in Sunday supplement reproduction for years – they have that too, but not on display. I leave feeling full.[7]

(5) Five o'clock. Too much Saturday: traffic so obviously on the way out. The wind is up to blow sounds of change – leaves starting to fall – against the traffic church bells peal and the market collapses noisily. People have the compact, satisfied look; couples are coupled as they walk, children drag, and dads are littered with household objects. The town has done its job and only the lost and lonely stand looking on.

These, then, are some responses to a place explored for the first time. The process of writing, like the process of painting, points up casual observation and releases new meanings. The painter John Hoyland[8] captured this well in his personal statement included in *The New Generation* exhibition at the Whitechapel Gallery:

The shapes and colours I paint and the significance I attach to them I cannot explain in any coherent way. The exploration of colour, mass, shape, is I believe, a self-exploration constantly varied and changing in nature: a reality made tangible on the painted surface.

Bedford by trail

If the analogy fits, then possibly a town trail is akin to painting by numbers, filling out a design drawn by another. The route is decided, the items for display are selected but ideally the route should still be an introduction to opportunity, suggestive rather than prescriptive. To obtain an impression of the Bedford that I should see, I followed one of two trails for the town developed by the Bedfordshire Association of Architects in 1975. Both walking trails are designed for two hours and meet the following description:

> We have therefore taken the opportunity during European Archi-
> tectural Heritage Year to produce voluntarily these architectural trails
> for use by the general public to draw attention to some of the buildings
> and areas in the towns and villages of Bedfordshire of which we can
> justly be proud. At the same time it is hoped that it will enable people to
> appreciate how new buildings built in our time and expressing our own
> age, can and should complement and worthily stand beside the archi-
> tecture that we have inherited.[9]

Trail 1, which I pursued, covered 24 points of architectural or historic interest in the town centre and immediate residential surroundings. Aside from encountering new places, it was my intention to measure my own reactions to the route against the checklist provided. Did it open the eye or did it blinker vision? Did it prompt personal discovery or merely link together some 'acceptable' view of architecture in place?

In the notes that follow, each stopping point (1–24) is noted, together with a selection of my own additions at appropriate points *en route*.

(1) High Street, (2) Howard Statue. Sunday morning – Sun day. Gusty coastal feel, landscape with figures, requests observation: a labelled tourist aims at the pensive John Howard (1726–90) and a file of clean-cut teenage hikers responds to the church bell. Shopping streets are superficially unin-teresting on Sunday, but detailing above can be seen (gold bull mounted wall clock). Which side should I be on? This side – that girl. Now to John Howard again, behind whom a utility local authority box with blue door – a statue loo? – more likely a market collector's store.

(3) Bridge, river and County Hotel.

> On the shallow, East of the 3rd pier of the Bridge stood the 'Stone-
> House' wherein BUNYAN imprisoned 1675–1676 wrote the first part
> of the 'Pilgrims's Progress'. 'As I slept I dreamed a dream.'

The County Hotel, standard First Class glass with a tower and raft look (the trail reports making use of the space between as an extension of the function of the buildings). From the bridge the river is clean and Midland, antiseptic save for immodest hoarding on the riverside club.

(4) Swan Hotel. Yes, but opposite the County Hotel, at an angle to the Swan is a Western Avenue mini-extravaganza in white, all tarted up, but all boarded up with nothing to show. 1930s boat freize, touches of Odeon and Tutankhamen. The 'BRIDGE HOTEL' sign, whitewashed over: splendid building, splendid site – no function, no trail mention (pause for Nairn sob), an earlier rear section with fire damage helped to explain.

(5) Gabled Terrace, St. Mary's Street. (6) College House, St. Mary's Square.

(7) The Small Cottages, 11 and 13 Cardington Road '. . . are two of the oldest houses in Bedford.' Yes, but what does mere age mean? – there's a milk bottle holder on the doorstep, *circa* 1975.

(8) St. Mary's House 'is interesting for its detailing' . . . and just had to be an architect's office. Could they have been responsible for what stands opposite? – 'Air Conditioned Prestige Offices, 16 200 sq. feet' tacked to a Victorian villa: empty, unkempt landscaping and a totally inappropriate use of the site. I have lost my way (always once on a trail). Interesting group of buildings – gothic–cum–Tudor revival (must be East Lodge) all in good shape: what was this area like a hundred years ago? More life I bet! Who lives here now?
'This is an opportunity to sit by the water and picnic'. Programmed relaxation – river activities, windblown trees, light playing, traffic distant. Every Sunday morning activity looking slightly conspiratorial! On a modest white plaque on the river wall, 'This Backwater of the River Great Ouse is where John Bunyan was Baptised *circa* 1650.' For an English autumn morning the light is good; the shadows are so much part of the place. Canoe trainees, working canal, well-kept early 20th century bandstand with signs of use, but possibly not by bands.

(9) The Suspension Bridge 1888. Mini–suspension bridge – view very trim and 'acceptable' riverside landscape. Flower beds and villas, what a lot of nice people – what a long way from the world – the niceness, the predictable tremble of views, patterns and events is disturbingly English. *But,* and it's a big 'but'; Gothic war memorial 1914–19 (with 1939–45 supplement) presumably George and the Dragon, but looks like Sarah Bernhardt dressed for Wagner atop an early Mondrian sculptural form with a Pre-Raphaelite monster at base – could well be all these things. Unsigned.
Embankment Hotel (how was this missed from the trail?): early 20th century arts and crafts, Chester–like half–timbering with contemporary windows and sign ironwork. Unfortunate modern entrance but fits well with a row which includes undistinguished but unobtrusive modern building.

On the river a men's pair, then a muscular women's eight. A small boy
wrestles with the gears of a guillotine lock, the boat enters and the expectant
crowds fade. 11 a.m. and the pace of life quickens – more walking, more
boats, more cars.

A block of stone in a flower bed:

> *This Cornerstone of Freedom Presented June 17,*
> *1948 at Bedford, Indiana, USA. The Stone City*
> *of the World.*

> GREAT BRITAIN

> To the People of Great Britain from the People
> of Indiana on the Occasion of the One Hundredth
> Anniversary of Indiana Limestone as a symbol of
> Friendship and Solidarity.

> Presented by the Indiana Limestone Company Inc.

I was there – Bedford, Indiana – the hole from which the Empire State
Building emerged. Rural Indiana speaks to the villas of Bedford. It would be
good to think of 1948, Truman-time Hoosiers explaining their right-wing
Middle American fears and claiming solidarity. But the Indiana Limestone
Company Inc. had its corporate tongue in cheek – Bedford, Australia, New
Zealand, Canada, South Africa – have they all got their Cornerstones of
Freedom too?

(10) The Museum, (11) The Castle Mound. Where? Over my shoulder
'Castle Close – Within These Grounds is the Ancient Mound Whereon once
Stood the Keep of Bedford Castle: Open to the Public.'

(12) Fire Station (1888), (13) Bunyan Meeting, (14) Congregational Church
(1774). Bunyan Museum – Closed Sundays. Sounds of lusty Free Church
singing behind closed doors. A scramble of buildings, new infill, car parks,
street furniture cut views, kids' paintings on walls (graffiti?) – interpret this,
distinguish this typical 'behind the High Streetscape' if you dare. I have got
lost again, but then find the Bunyan Meeting, quite an event with glass doors
allowing a facsimile of 'Stars on Sunday' to the service beyond.

(15) St Cuthberts. The Established, but deserted, church: a Victorian–
Norman keep of a place, little loved and unadaptable, its only market a label
'25p Off Recommended Price' stuck for the self-adhesive fun to the locked
door. The local shops do better and know it: a butcher's shop has three
butcher's straw hats (Bedford straw no doubt!) and two crossed meat cleavers
on the weekend slab. Again the town falls away – the inner town, no man's
land. An arrow offers Car Park, 840 Cars and the eye follows to the Merton
Centre, brick and grey-tiled mansard block from anywhere – '93 000 Sq. Ft.

Office, 4 Shops to be Let.' Beyond, T. H. Smith and Sons, Corn Merchants, look on amazed.

More disasters in Grove Place. Gladstone Villas now house an accountant and someone who calls the house 'Moel Siabod'. Even if you can get two personal names out of that would Gladstone understand the joke?

(16) The Grove 'on the other hand retains the comfortable residential feel.' A Thirties house approached from the rear on the corner – modest excitement as a semicircular stair window is sighted, but oh, 36A is a disaster; senses twitch with the breeze of a Sunday roast. Improved residential with asters, butterflies and roses filling it out.

(17) Rothsay Garden '. . . a different scale of domestic building'. Edwardian mansions and good planting – subdivision must hover threateningly in the background, but there could be money here to keep it. A very pleasant, urban, place; envy is the chief emotion – urban, affluent and closeted. Scarcely hidden on the high road: Heron House, all concrete stilts and out of scale, another office block which contains the Department of the Environment's Eastern Region Controller. As is not the case at Marsham Street (the Department's London headquarters) the trees and the address soften this blow.

On to Goldington Road, corporate places, new blocks, the building line is gone and the road begins to rule – May House, Zurich House, Graylaw House (HM Customs and Excise, Value Added Tax Mon.–Fri. 9 a.m.–4 p.m.)

(18) Graylaw House (1972, architect Peter Smith Associates). One of the better new buildings in Bedford and the subject of an architectural award; the general scale, height, materials and window form are all appropriate for the area. It is tight, urban, attractive colour and texture, wears well, poorly landscaped and, compared with others (like the tat of the Norwich Union Insurance Group next door), deserves an award. But what was here before, and why string such pearls on the road? Roads now rule, who cares what fringes them? – they determine life here. A young girl sits on the car park wall, drawing or writing and staring at the insurance blocks beyond; I hope that she *is* writing a letter, it's not worth the drawing. A lot of people seem to be muttering as they walk today . . . I am doing it too.

(19) Houses in Kimbolton Road. More acceptable dwellings but what of the hospital shambles hidden from view – why are we trailing up here? – and here – 'the walk is long but look for the varying details of bay windows . . .' Fantastic! Look at neat trimmed gardens, expensive cars, the rustle of Sunday brunch . . . and look at him, fronting 'Bedford School – Private Grounds', a blond lad of ten leaning on the gate awaiting parent or guardian. Sad in grey

suit, grey socks, black shoes and tie – a nation trained to mourn – minute against the backdrop of green field. He is picked up by a male in an L-registration white estate car. Unhappy elements of a story unfold; Beckett, Pinter and Co., have much to answer for.

(20) Bedford Park. Expensive green, very North American in space use, excellent of its kind; a table set for Edwardian lunch – who washes up? Opens a good vista of Bedford School, English brick Gothic with modest Rathaus trim fronted by a hop field of rugger posts. Two trim young vicars roll past in a Mercedes, it's all becoming too much like the movies' English Sunday. Church is out and one-day suits walk their way through the sprinkle of leaves.

(21) Robinson Pool, (22) De Parys Avenue, (23) St Peters, (24) Offices of Redpath Dorman Long. At the Broadway, the John Bunyan statue and Parish Church on an island of green moving at Sunday pace, are surrounded by traffic and the odd nothing-to-do spectator. I never quite finish any route.

Place experiences

Although making no claims for literary merit or acute observation, the record of freely derived and trail-guided experiences above represents something comparatively rare in academic, or even popular, literature. I would suggest that few people have taken the opportunity to experience a place with that purpose alone in mind. The experience of places is, and for long has been, a formalised process, called amongst other things tourism, touring and visiting historic towns. There are necessary preparations and post-visit activities which make the experience itself, at most, only part of the process. Guidebooks and souvenirs package places, the camera roves for the record and traffic, parking and accommodation schedules organise the activity.

How does the programmed presentation of place differ from the more open experiences which I have hinted at above? Reviewing my own notes on Bedford, ten terms emerge as underlying the observations. First, there is the context of *weather* and *time*. Second is experience, characterised by *feel* and by *activity*. Third comes action, suggested by *navigation* and *movement*. Fourth, the process of organising *information* using *selectivity*. Finally decision, involving *judgements* and *response*.

Interestingly, although a specific place with a built-form and man-made green environment provided the setting, and would form the focus for any formal interpretation of the place, these elements do not figure in the list above. Was this, I asked, an over-reaction against a professional interest in 'the environment', an unconscious attempt to play down the role of built-form and physical features in order to make my point? I think not, although couplings such as 'man–environment' encourage the academic to dwell on the resolution of a supposed dichotomy, day to day life for most of us does not involve the erection of such hurdles. As Philip Wagner[10] has observed:

> [The] individual and his environment, equally physical and social . . . are in fact one. A person and his context and actions, as well as a people and its environment, can best be seen as indivisible.

In experiencing this indivisibility, what elements come to the fore? The *weather* and the complexities of micro-climate are more important to the individual than we like to admit. Experiencing place is to observe and participate in the detail of location and movement which offer sun, shade, cooling gusts, the flicker of tree shadows.

The experience of *time* in place is something which Lynch[11] had addressed, even proposing some management techniques for emphasising or promoting the temporal aspects of urban places. In comparison with the physical aspects of place, time seems far more subtle in its influence on human behaviour. The ebbs and flows of the bus station were movements directed by a complex mechanism of bus arrivals and departures, shop capacity and pedestrian traffic. Time spent in the museum or gallery was influenced by those around me, the arrangement of exhibits and the level of attraction involved. Daily and weekly rhythms have a profound effect on the sense of a place – Bedford at midday Saturday and at midday Sunday was two different places.

In terms of experience, the oft-quoted (even by this writer) surrogates of perception and latterly 'cognition', seem all too hollow when one's total experience of place is assessed. While the visual sense may be the best developed, we certainly supplement it with hearing (especially in terms of melody or sudden shifts in volume), with smell (especially in terms of food or smoke), and even with touch (especially paved surfaces or in crowds). Moreover there *is* a sixth sense, the *feel* of a place, largely set on a continuum ranging from fear to pleasure, or as Faber[12] puts it, stress and creativity. The level and form of human *activity* have a major role to play here. Towns and cities are not full of neat, Cullenesque 'townscapes' where only decorative people dwell; they are teeming with activity, never to be captured by statistical analysis, film . . . or writer. The urban context is largely a human one and our feelings about places reflect more the glances or gazes of other people's eyes, snatches of conversation, clothes and appendages, than the built-form background.

The same is true of our primary aim in place, that of *navigation*. Our need to achieve goals within urban space means that much effort, conscious or otherwise, is directed towards getting from A to B. Navigation techniques at the local scale depend not on the logic of measured distance but on the shifting patterns of stimuli, built-form at base, human and unpredictable in detail. Pace and style of *movement* are affected similarly – halts, side-steps, pauses for observation are more common than the designer's favourite, meeting places. Although many patterns may be predicted and tested by, for example, introducing a street event into a square, it is the local subtleties and arrangement of such predictable patterns that provide a key element in the silent identity of a place.

The experience of *information* is a more obvious characteristic of place. The notes above repeat selections from written information in the Bedford environment, but lettered and other graphic messages and symbols are not the only forms of information available, even if they are most evident. Buildings do offer their own language, especially with regard to function and access. Aside from the obvious use of language, as in the response to a query, people in places offer both individual and group messages for the taking – in clothing, facial expression, speed of movement and clustering. There is far too much information for us to digest in most urban places.

In Britain – if we are British – we know the code and use considerable *selectivity* in obtaining the information which we need, either for direct action or for more cerebral or emotional pleasures. In a novel environment, especially if a different culture, we often lack the code and have to establish a gradual selectivity over a large range of potential inputs: a process of place learning. Recent personal experiences of Venice were instructive in this respect: covert street signs indicating the types of area, canal bridge/church/ square configurations as community centres, native clothing, smells more than in most other places I have visited – all these are key information sources in Venice.

Having experienced a place, we tend to involve ourselves in a continual series of *judgements* and *responses* to people, places and situations. We put ourselves in the place, attempting either to share its values or to impose our own. The judgements may be minute and restrained, or full of impact (as with the football crowd's urban progress), but each user of place contributes to the nature of that place.

All of this may seem abundantly obvious – self-evident observations on a process of urban place use known to all – but is this the case? I would suggest that relatively few urban place-users consider their experiences and, while familiarity breeds either contempt or smug satisfaction with the known home town, are we equipped to experience the riches which any novel environment has to offer?

In academic terms the impressionistic account of person in place cuts through so many neatly bounded disciplines (e.g. sociology, psychology, proxemics, man-environment studies, behavioural geography, urban design and the rest) that we tend to lose sight of the obvious. The impact of these disciplinary stances on most people's experiences of place has been slight, but I am struck by the similarities between the fragmentation and potential for de-humanisation of place which the extended pursuit of many of these disciplines has involved and a recent observation by the photographer

Tom Picton[13] on the subject of architectural photography. He compares the product of 1930s photographers working for *The Architectural Review* with the work of other photographers who contributed a major photo feature, 'Manplan', to the same journal in the 1960s:

> The decay of the architectural photograph into sterile perfection is part of a much larger movement in our culture which everybody feels and nobody can explain. Laszlo Moholy-Nagy looked at people in Brighton – even from a distance – enjoying themselves at the seaside. Patrick Ward, Ian Berry and other photographers in the 'Manplan' issues of the AR showed these same people as victims of an uncaring society. The '30s' were perilous but the photographs were optimistic. Ten years ago, in 'Manplan', people looked miserable or, perhaps more exactly, catatonic, even on holiday. We no longer write about the common people, but 'Manplan' photographed them, mostly without affection.

The key word is 'affection'. In our search for an understanding of place experience, a search which has led to the analysis of perceptions, behaviours, histories and spatial patterns, we have continually excluded the transitory or deep-seated emotional relationships which may link person and place for a second or for a lifetime. As in architectural photography, the forced message evokes an intellectual rather than emotional response.[14] So much of what we have come to regard as 'useful' literature on place excludes feeling as unbecoming the academic setting.

Not surprisingly, the 'feeling gap' has been transplanted and, I believe, has broadened in the latest developments of urban interpretation. The origins of the amenity society movement, now dominated by the Civic Trust, certainly reflected feeling with regard to the loss of valued places and buildings. Multiple stores, multiple building designs, and multiple fashions so homogenised our urban centres that individual characteristics and values of place were being lost. Amenity interests were often accused of being over-emotional in the face of impending loss and of failing to make rational judgements concerning the 'proper' use of resources. After 20 years or so of amenity interest, however, we have perhaps come full circle. Conservation has become formalised and it ensures that Britain remains an effective respository of the past.

In the concluding section of this essay, I want to suggest three avenues of exploration which I believe can be followed in the confidence that students will emerge with a realisation that 'feeling' in place is a valid and valued experience.

Provoking place experiences

The process of ensuring that place experience is on the personal agenda of more people, in more places, is a complex one and only suggestions are possible here. Returning to Picton's observations (p. 23), I take leave to doubt the assertion that 'everybody' feels this decay into 'sterile perfection', but I believe that the characteristic can be identified in geography and in geographical teaching. Geography is a discipline with strong academic traditions and an established role in the British school curriculum. How does geography teaching provoke place experience? Over the past two decades there has been a decline in place-related work in favour of space-related studies which often exhibit 'sterile perfection'. So far has geography moved from the displayed affection for people and place, that the occasional descriptive essay on place becomes something of a novelty. Reviewing *St. Croix Border Country*,[15] Peirce Lewis[16] describes it as 'subversive because it says cunningly, obliquely, and with high style that it is about time professional geographers in the United States started using their eyes again.' The book narrates a two–day car journey east of Minnesota's Twin Cities with roadside observations, of which the following is both typical, and significant:[17]

> Over on the south side of the lake, in a bowl in those morainic hills, a local entrepreneur is building an auto racing track. For several seasons now, the hills have rung to the sound of big-bore stockers churning around an oval dirt track. The competitors, for the most part, are also owners of the cars, and tend to be young men from the farms and nearby small towns. But now, in a move that almost reeks of socio-logical significance, the track owner has decided to build a paved road circuit and operate under the aegis of the Sports Car Club of America. It is the triumph of the urban sophisticate, with his delicate little European thoroughbred, over the rude country boy, whose heavy, hard–charging piece of blown Detroit iron may go fast enough, but definitely lacks class. The classic description of this process is Tom Wolfe's essay, 'The Last American hero' in his volume, *The Kandy-kolored Tangerine-flake Streamline Baby*.

As Swain and Mather note in a footnote reference to Wolfe's book,[18] 'this contains some amazingly acute perceptions which have not yet occurred to cultural geographers.' Ten years on, 'the new journalism' of which Tom Wolfe was such an essential part in the USA has come and gone[19] and, like *St. Croix Border Country*, has left little mark on geography. Like James Agee,[20] Wolfe has managed to capture the sense of places inhabited by people, and in a style which itself conveys something of the pattern of contemporary America. More important, the writings of Agee, Swain and Mather, and of Wolfe make one want to go and experience for oneself.

To take an earlier aside, why a mark on geography? Because I believe there is, and rightly, a popular expectation that geography has cornered the market when it comes to knowledge of place, and that the discipline has a duty to society to give of that knowledge and to encourage exploration and environmental experience. As a subject it holds its own in schools and colleges, at least partly because society believes that 'place' is being taught. If not by geographers, then who else has the responsibility or the knowledge? True, other professions and disciplines have a part to play, and the literature of environmental education in schools suggests that as much place-based teaching may occur in the history class as in geography. It is in schools and other places of education, with their captive audiences, that we are likely to see any positive expansion in place experience occurring. Although largely ignored by recent literature on the subject, educational town trails have been used in an innovative and challenging manner, and personal experience suggests that the preparation of a local area trail by students of any age can be a valuable place-exploring experience. It was from this knowledge that the first technique discussed below was derived.

Sensing the environment. One modest, yet direct, attempt to develop sensory awareness of place has been the production of a unit for the Schools Council entitled *Art and the Built Environment.*[21] The Schools Council project briefly notes the initial aims of the project as being 'to enlarge the students' environmental perception and enable them to develop a "feel" for the built environment [and] to enhance their capacity for discrimination and their competence in visual appraisal of the built environment.' We were commissioned to prepare the unit as an issue of the *Bulletin of Environmental Education* in an attempt to stimulate interest in experiencing the environment amongst trial schools using the various project materials.

In developing the unit we drew on the work of a number of educators and designers who had prepared student and participatory materials, largely in the United States. The work of Henry Sanoff[22] was influential, as was that of Saul Wurman[23] and the design-related studies undertaken by Lawrence Halprin,[24] an American landscape architect. I had previously reported on 'sensory walks' which had been developed for community and student groups in Birmingham and, with the rapid growth in 'town trails', the multi-sense exploratory trail or walk was a development which I had hoped would prove attractive to trail producers.[25]

There is, however, something rather precious about asking the trail user to stop and smell a flower at point A, or to switch from analysing the structure of a Georgian doorway to feeling that same doorway. We decided that in the school context at least, the best approach to widening the sensory experience of place was to set the activity in the form of a game which involved the students at all stages of its development. Figure 2.2 shows the 21 symbols which form the basis of the game. These can be modified or supplemented as

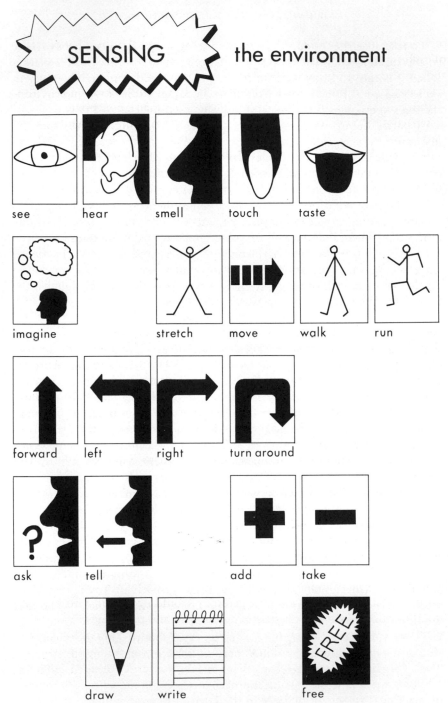

Figure 2.2 'Sensing the environment' game symbols.

the participants prepare sets of cards for their own use. The number of cards of each type depends on the area to be used, the time allowed for the activity and, in the school context, the skills which the teacher wishes to develop. Each participant prepares his or her own set of cards, shuffles the pack and advances into the immediate environment. Using an appropriate time or distance interval the top card is placed on the bottom of the pack and the next card is 'obeyed'.

Most symbols are self-explanatory. *See, hear, smell, touch* and *taste* invite emphasis on one sensory experience, while *imagine* invites something rather more than experience. *Stretch, move, walk, run,* in relation to *forward, left, right* and *turn around* provide the movement and navigation elements. *Add* and *take* imply the addition or subtraction of something from the immediate environment and, although it is possible to undertake these instructions legally, some school groups have inevitably run into trouble with these headings! *Ask* is fairly easy but *tell* offers the student (or adult) more of a challenge. *Draw* and *write* are the two basic recording skills and these can be supplemented with film, collect, rub, measure etc. if a product of the activity is required. The *free* card allows flexibility or other activities not listed above. Given an enterprising teacher and uninhibited children or students, this basic game can be adapted to a variety of situations and works well. In at least one school, parent groups have also become involved.

'Sensing the environment' is clearly only a beginning. Halprin developed 'scores' by which a series of directions and activities bring the user into novel experiences of the home town – these were used in his work[26] for Everett, Washington. Halprin[27] also pioneered the use of notation systems in design and design analysis. Such graphic methods, of which a range have been developed by professionals seemingly as far apart as doctors, ballet choreographers, landscape architects and psychologists, allow the observer to reproduce in a notebook something of the activity and vigour of urban life. Clearly photography must be a supplement if a detailed analysis is to be effective, but even a basic notation system such as that used by Halprin can be used by students in order that they focus on the urban experience of others, thus reflecting on their own.

Environmental autobiography. A major purpose of 'Sensing the environment' was to place users in situations which they would not usually experience. Much has been written[28] about our stereotyped and patterned responses to environment, to each other and to culture. Most current literature on place and places serves to reinforce, rather than cause reappraisal of these stereotypes. 'Sensing the environment' works largely because it places the burden of experiencing and finding meaning squarely on the individual. No one else's system of mirrors intervenes.

Kenneth Helphand[29] has recently reported on his experiences in generating environmental autobiographies from his design students in Indiana and

Oregon, and here, too, we can see an opportunity for individuals to regain contact with environmental experience, their own experiences of the past. Through an extended period of research and studio presentation, students are encouraged to tap their memories and inner resources, to recapture or perhaps revisit environments from their past lives. In viewing the rich material generated by this project, one is struck by the depth of our own past environmental experience which we hide or ignore when observing or making decisions in the present. The environmental autobiography, like other less graphic self-exploratory techniques, serves to strip away, or at least put for reappraisal, many of the patterns of observation and structures for information which we tend to adopt in the complexity of middle and later life. Like the sensory walk experience, environmental autobiographies need not be just feeling experiences, as Helphand notes:

> In the course of doing environmental autobiographies individuals sub-
> jectively communicate their past environmental experiences. They are
> then analyzed and synthesized in informal and formal presentations.
> The process of presenting aspects of one's environmental history acts as
> a transition between 'intuitive' and 'conceptual' understanding. As one
> might expect often the intuitive experience is a validation of theoretical
> concepts, however the process of personal induction of theory cannot be
> underestimated. Instead of being taught a set of ideas, conceptual under-
> standing is discovered latent in individual experience and the shared
> experience of others. The discovery process both validates and helps im-
> print ideas, while equally questioning their validity and applicability.[30]

The mind searching provoked by the environmental autobiography stirs the imagination and improves receptiveness to environmental experiences. I believe that the sharpening of environmental awareness is better achieved through such tools, though the quality of an individual experience of place can undoubtedly be enhanced by the availability of interpretive material.

Opportunities in interpretation. I turn to the matter of interpretation with some misgivings, especially in the light of the rapid growth of literature in this field. It was a fear that the literature was pursuing too narrow a goal which prompted this essay, but it would be incorrect to suggest that interpretive efforts in general should be discarded or rejected. Personal *experience* of place should be paramount, but the most common and accessible information on place is usually provided in printed form – in leaflets, wayside markers, descriptive labels and trail brochures. These are all tools for the interpretation of place by local authorities and citizen groups.

The development of interpretation as an organised professional activity has, until recently, been largely American and has taken place in rural rather than urban areas, in recreational rather than everyday environments. The classic document is Tilden's essay[31] from which principles of interpretation have been drawn by various authors.[32] The two principles which I regard as the most important in the British urban context are that, first, interpretation must relate what is being displayed or described to something within the personality or experience of the individual and, second, that the primary purpose of interpretation is provocation rather than instruction.

Although there is an increasing range of information on interpretive methods,[33] these two principles are all too easily neglected. It is, in fact, very difficult for the expert who wishes to communicate something about a place to avoid instructing and so wrapping up his presentation in professional terms as to exclude the opportunity for personal contact with the eventual user. A section from Percival's manual under the heading *Stimulate thought and further exploration* is indicative:

> For the most part, though, interpretation is probably at its most stimulating when it pinpoints and analyses those urban qualities which though undeniably attractive have often defied analysis before. One example is the old building, wholly devoid of architectural polish, which still has charm. Why? It has evolved organically over the centuries, alterations and additions having been sympathetically undertaken in local materials to produce a harmonious whole.[34]

If such an analysis leaks into interpretive literature there is every danger that 'analysis' will become 'instruction' and that the opportunity to feel and experience an undiscovered environment is again removed.

The major task for interpretation in the urban context is the effective guidance of residents and visitors towards opportunities for discovery and experience. An increasing number of booklets achieve this purpose and amongst these I would cite the Shire series[35] and various documents produced by local authority planning departments.[36] In Bedford, a most informative paper by Hassall and Baker[37] on the evolution of the town was readily available at the museum. A sound historical report with maps, this booklet provided the visitor with a background to the town's past but did not foreclose the experiences of past environments which might be gathered from personal exploration.

On-site interpretation of British towns is at a frustrating halfway stage at present. Most towns of appreciable size have a Tourist Office, but too often, as in Bedford, it is closed when needed or, if open, fails to offer any co-ordinated pattern of local opportunities. If there were no tourist offices signposted or trail facilities, the visitor would be forced to explore and experience. If all the facilities were of a high order then most visitors might take only a programmed experience of place, or they might use the pro-grammed experience as a satisfactory beginning to their own exploration. With fragmentary facilities, frustration often tends to overwhelm experience of place. Even after considering my own response to Bedford, I remain convinced that there must be more investment in interpretive *planning* in urban areas. At its most basic, this means that information is available where people expect it, that the routes offered go where signposted, that the facilities offered are open to the public, and that significant features for which a town is known can be examined. In short it means *making a place observable*.[38] Many of the basic techniques are already well known, such as information centres, marked routes, trail leaflets, interpretive displays, open factories and workshops, and trained guides. All too often, however, there is little co-ordination of facilities or common purpose in presentation. Designing the plan brief for interpretation of a place seems to be an almost totally neglected activity. In addition, I am convinced that most urban places could offer a much wider range of interpretive opportunities than is presently the case. Towns can give much more of themselves and invite the visitor to participate in a much richer experience, as may be seen from two recent American examples.

First, with the assistance of William Carney,[39] the town of Leverett (Massachusetts) prepared a booklet of comments derived from the towns-people, which covered aspects of the local environment as they saw them. Carney, now working under the title *Landscript,* has applied this technique in other US urban areas. Although the editorial function is crucial, and should be as neutral as possible, the technique offers the resident and visitor a series of unofficial insights into the place as seen by its people. Of course, one can actually talk to people in a place and obtain something of the same result. If such *vox pop* statements were as readily pursued as are town trails, then the built-form dominance of place images would undoubtedly be dented, as people talk more readily of people than of buildings.

The dominance of built form in urban interpretation can also be altered if town trails and other architectural descriptions are set in a new mode. Such was offered by the second example, the Cooper-Hewitt Museum's catalogue to its *Immovable Objects Exhibition* in New York City.[40] Taking the form of a fifty-cent tabloid newspaper, the catalogue revealed that the 'exhibition' was, in reality, 'an outdoor exhibition about city design on view throughout Lower Manhattan from Battery Park to Brooklyn Bridge.' The catalogue served as a 'trail' linking buildings, some of which had local exhibitions, but also offered provocative historical and political comment, services that were available and experiences that could be enjoyed. It also included discussion of conservation issues, and interviews with designers and other influentials. The main lessons to learn from the Cooper-Hewitt catalogue, and the route which it identified, are that a document can, and should, contain a variety of interpretations of a place and that the features linked in a trail should be capable of use and exploration rather than merely of observation.

Implications

After a recent seminar discussion at a College of Education, a mature student asked, after a convoluted exploration of recently digested academic books, how he could possibly encourage his future students to experience places when he could not experience them himself. Although destined for environmental studies and geography teaching, this student could find no inner resource which encouraged him to believe that he could learn from the environment and stimulate others to do so.

Although it is unlikely that this student will read the notes above, it is to him, and to the many like him, that the words are directed. Over the past 20 years, or perhaps longer, many people have lost touch with the ability to sense place. I have attached some blame for this to misdirection in geographical education, but this error is small compared with the impact of rapid place by-passing travel, media stereotyping, packaged vacation experiences and standardised design. Like many others, I have seen the potential in urban interpretation for the rediscovery of place by tourists, visitors, and possibly residents, but recent developments in interpretation have led me to doubt whether this new profession will be provocative rather than instructive.

In the opening notes on my experiences in a novel environment I have jogged my own mind, and hopefully yours, with regard to the fantastic range of opportunities inherent in experiencing any place. Naturally, Bedford is unique and has many assets which I have not mentioned, but it is also typical of our market towns, places available to all. In reporting my experiences on the trail I have indicated that any trail *can be* the vehicle for unstructured exploration but I have considerable doubts as to whether these historical and architectural guides will provoke, rather than merely instruct.

I have suggested that new techniques are available for use in preparing students, the future public, to feel places rather than just to learn from them. I have indicated that there are also techniques available for broadening, rather than narrowing, the practice of interpretation.

In conclusion, I return to Picton's observation[41] that 'the decay of the architectural photograph into sterile perfection is part of a much larger movement in our culture which everybody feels and nobody can explain.' Though I am unhappy with absolutes of 'everybody' and 'nobody', this view is one which, I believe, does apply to our experience of place, and therefore to professional photographs of it. In searching for acceptable methods for educating about, interpreting, and designing place, we have erred on the side of 'sterile perfection'. Security of tested method in teaching, interpreting and designing has replaced method inspired by the individual's experience of place.

In this essay I have suggested ways in which the educator and the interpreter can avoid the path to sterile perfection. Elsewhere there have been a number of suggestions aimed specifically at the urban designer. [42] Although each individual can take steps to re-establish personal links with an environment for which feelings have been numbed, it will require the support of the professions identified above if our society is to regain contact with the environment of feeling which we have mislaid.

Acknowledgements

For a variety of reasons, I should like to acknowledge the assistance of James Agee, Ian Bentley, Blondie, Hugh Dawson-Walker, John Gold, Tom Harrisson, Ian Nairn and Tom Wolfe in the preparation of this essay.

Notes

1 Goodey, B. 1977. *Interpreting the Conserved Environment: Issues in Planning and Architecture*. Working Paper 29, Department of Town Planning, Oxford Polytechnic.
  Goodey, B. 1978. Where we're at: interpreting the urban environment. *Urban Design Forum* **1**, 28–34.
2 Percival, A. 1979. *Understanding Our Surroundings: A Manual of Urban Interpretation*. London: Civic Trust.
3 See Goodey, B. 1975. *Appreciation of the Aesthetics of the Environment: The Experience of Urban Trails*. CCC/DC (75)6, Committee for Cultural Co-operation, Council of Europe, Strasbourg.
  Goodey, B. 1975. *Sensory Walks*. Working Paper 33, Centre for Urban and Regional Studies, University of Birmingham.
  Goodey, B. 1975. Towards new perceptions of the environment: using the town trail. *Bull. Environ. Educ.* **51**, 9–16. Revised for D. Appleyard (ed.) 1979. *The Conservation of European Cities*, 282–93. Cambridge, Mass.: MIT Press.
4 For extensive topographical description see Banham, R. 1971. *Los Angeles: The Architecture of Four Ecologies*. London: Allen Lane.
  For concise observations, see shorter essays such as Banham, R. 1977. Sundae painters. In *Arts in Society*, P. Barker (ed.), 159–63. London: Fontana/Collins.
5 The original Mass Observation report is in Madge, C. and T. Harrisson (eds) 1939. *Britain by Mass Observation*. London: Penguin Special.
  A detailed bibliography of Mass Observation publications is included in Harrisson's last book: Harrisson, T. 1978. *Living Through the Blitz*. London: Penguin.
6 Cooper, M. C. 1978. A tale of two spaces: contrasting lives of a court and plaza in Minneapolis. *Am. Inst. Arch.* **67**, 34–40.
7 The disparity between the displays in the museum and in the art gallery must largely result from funding. The Cecil Higgins Art Gallery, named after a local brewing benefactor, is exceptionally well endowed. Its claim to being 'renowned internationally for its collection of English and Continental ceramics and glass, English water-colours, prints, furniture, *objets d'art,* costume and lace,' cannot be disputed. The interest from the endowment has been used to build an excellent collection of water-colours and to purchase a large proportion of the Handley-

Reed Collection of Victorian and Edwardian decorative arts. Information and interpretive displays in the gallery are of a very high quality.

8 Hoyland, J. 1964. 'Statement' in The New Generation Exhibition at Whitechapel Art Gallery, reprinted in A. Brighton and L. Morris (eds) 1977. *Towards Another Picture: An Anthology of Writings by Artists Working in Britain 1945–1977.* Nottingham: Midland Group.

9 Bedfordshire Association of Architects 1975. *Architectural Heritage: Bedfordshire Trails.* Set of six leaflets produced by the Association in Bedford.

10 Wagner, P. 1972. *Environments and Peoples,* 100. Englewood Cliffs, NJ: Prentice-Hall.

11 Lynch, K. 1972. *What Time is This Place?* Cambridge, Mass.: MIT Press.
See also the essay by Parkes, D. and N. Thrift 1978. Putting time in its place. In *Making Sense of Time,* T. Carlstein, D. Parkes and N. Thrift (eds), 119–29. New York: John Wiley/Halsted Press.

12 Faber, S. M. 1966. Quality of living – stress and creativity. In *Future Environments of North America,* F. F. Darling and J. P. Milton (eds), 347–8. New York: Natural History Press.

13 Picton, T. 1979. The craven image or the apotheosis of the architectural photograph. *Arch. J.* 25 July, 175–90.

14 ibid., 187.

15 Swain, H. and C. Mather 1968. *St. Croix Border Country.* Prescott, Wisconsin: Trimbelle Press.

16 The review of *St. Croix Border Country* appeared in the *Geog Rev.* **61,** 159–61, 1971.

17 Swain and Mather, op cit., 29.

18 Wolfe, T. 1965. *The Kandy-kolored Tangerine-flake Streamline Baby.* New York: Noonday Press (and subsequent paperback editions).

19 The school of 'new journalists' has been reviewed by Wolfe, T. and E. W. Johnson (eds) 1975. *The New Journalism.* London: Picador/Pan.
The style survives and has been developed by Wolfe, T. 1979. *The Right Stuff.* London: Jonathan Cape, an original and penetrating discussion of the US space programme.

20 Agee, J. and W. Evans 1939. *Let Us Now Praise Famous Men.* New York: Ballantine (since reprinted with additions).

21 Goodey, B. and W. Menzies 1977. Sensing the environment. *Bull. Environ. Educ.* **72.** The Schools Council 'Art and the built environment' Project was directed by Colin Ward and Eileen Adams at the Education Unit of the Town and Country Planning Association; a book-length assessment of the project is to appear from the Schools Council. Eileen Adams is now directing an extension of the project from the Royal College of Art (1980 on).

22 See, for example, Sanoff, H. 1975. *Seeing the Environment: An Advocacy Approach.* Raleigh, NC; and Sanoff, H. (ed.) 1976. *Learning Environments.* Asheville Environmental Workbook, Raleigh, NC: Community Development Group, School of Design, North Carolina State University.

23 Wurman, R. S. 1971. *Making the City Observable.* Cambridge, Mass.: MIT Press.
Wurman, R. S. 1972. *Yellow Pages of Learning Resources.* Cambridge, Mass.: MIT Press.
Wurman was also involved in editing (1976) *An American City: The Architecture of Information (Conference Programme).* Washington, DC: American Institute of Architects.

24 Halprin, L. 1969. *The RSVP Cycles: Creative Processes in the Human Environment.* New York: Braziller.

25 See note 3 above.

26 Lawrence Halprin and Associates 1970. *Everett*. Report by Lawrence Halprin Associates for the City of Everett.
27 Halprin, L. 1969. op. cit.
28 As in Berger, J. 1974. *Ways of Seeing*. London: BBC/Penguin. Also Sontag, S. 1978. *On Photography*. New York: Delta.
29 Helphand, K. I. 1979. *Environmental Autobiography*. Paper to International Conference on Environmental Psychology, University of Surrey, Guildford.
30 ibid., 2.
31 Tilden, F. 1967. *Interpreting Our Heritage*. Chapel Hill, NC: University of North Carolina Press, rev. edn.
32 Sharpe, G. W. (ed.) 1976. *Interpreting the Environment*. New York: Wiley.
33 Basic handbooks include:
    Aldridge, D. and K. Pennyfather 1975. *Guide to Countryside Interpretation*. Vol. 1: *Principles of Countryside Interpretation and Interpretive Planning*. Vol. 2: *Interpretive Media and Facilities*. Edinburgh: HMSO for the Countryside Commission and the Countryside Commission for Scotland.
    Alderson, W. T. and S. P. Low 1976. *Interpretation of Historic Sites*. Nashville, Tenn.: American Association for State and Local History.
34 Percival, A. op. cit., 17.
35 For example, Haddon, J. 1970. *Discovering Towns*. Princes Risborough, Bucks: Shire Publications.
36 For example:
    Bott, O. and R. Williams 1975. *Man's Imprint on Cheshire*. Chester: Cheshire County Council Planning Department.
    Leicester County Council 1975. *The Local Tradition*. Leicester: Environmental Services Section, Leicester County Council.
    Tillyard, R. 1979. *Nottinghamshire's Heritage: A Strategy for Its Interpretation*. Nottingham: Nottinghamshire County Council and Countryside Commission.
37 Hassall, J. and D. Baker 1974. Bedford: aspects of town origins and development. *Bedfordshire Historic Environment* **2**. Bedford: Bedfordshire County Council Planning Department and North Bedfordshire Borough Council Museum (originally in *Beds Archaeol. J.* **9**).
38 This process is well-documented in Fondersmith, J. 1974. *Making Washington Observable: A Program for Civic Education and Visitor Orientation*. Washington, DC: District of Columbia Office of Planning and Management.
39 Carney, W. 1973. *Where We Stand: A Report on Leverett's Planning Process*. Leverett, Mass.: Leverett Conservation Commission and Planning Board.
40 West, S. (ed.) 1975. *Immovable Objects Exhibition*. New York: Cooper-Hewitt Museum.
41 Picton, T. op. cit.
42 As, for example, in:
    Carr, S. 1967. The city of the mind. In *Environment for Man: The Next Fifty Years*, W. R. Ewald Jr. (ed.), 197–321. Bloomington, Ind.: Indiana University Press.
    Lynch, K. 1976. *Managing the Sense of a Region*. Cambridge, Mass.: MIT Press.
    Goodey, B. 1979. Going to town in the 1980s: towards a more human experience of commercial space. *Built Environ.* **5**, 27–36.

3 *Filming the Fens: a visual interpretation of regional character*

JACQUELIN BURGESS

In 1977 I was given the opportunity to participate, as consultant and co-researcher, in making a television documentary about regional character. Entitled *A Sense of Place: The Fens,* it was intended as a pilot programme for a series of regional films to be shown by BBC Television at peak evening viewing. It represented a departure from the traditional format of regional documentaries that were made for television during the 1970s. Such films as *The Making of the English Landscape, A Bird's Eye View* and *A Writer's Notebook: The Pennines* had depended on an expert commentator, either literary or academic, to make the necessary interpretations about the landscape and to give credence to the subject matter.[1] The Poet Laureate, Sir John Betjeman, eulogised from the cockpit of a helicopter, while W. G. Hoskins extemporised from Lake District mountainside or Norfolk marsh. By contrast, our film was made without any reliance on an expert reading a commentary or appearing in front of the camera, the intention being to explore and elucidate by means of film those values that the inhabitants of a region themselves attribute to their landscape, and to capture the particular, special and significant features of the place as local people experience them. 'Reading the landscape is a humane art, unrestricted to any profession, unbounded by any field, unlimited in its challenges and pleasures.'[2] In this essay I want to describe some of the challenges and pleasures as well as the difficulties that were encountered in capturing the sense of place.

Finding the story

A good documentary film has been described as one 'which earns the shock of recognition from a mass audience'[3] and may take one of two approaches to its subject matter. It may seek to provide an objective and impartial account of things, people or activities, or, alternatively, it may be a personal film which deliberately embodies the perceptions, emotions, interests and biases of the director. Personal documentaries are often more 'concerned with people rather than things, with the "why" of life rather than the "how" of it.'[4] John Schlesinger, the film director who made a number of personal documentaries for British television in the early 1960s, describes them as 'catching spontaneously the essense of what we had seen'.[5] This comment highlights the most significant point about the content and style of these films. Their subject matter is that of everyday life: 'the drama on the doorstep, the drama of the ordinary', and the style is that *cinéma vérité*.[6] Ordinary people going about their normal lives provide the content for the film, with no predetermined scripts, actors or shooting schedules. *Cinéma vérité* asks nothing of people beyond a willingness to be filmed, and the camera is used to compile visual notes which are later edited to produce the story of the film. Devices such as music or intermittent commentary have given way in recent years to a reliance on 'wildtrack sound', in which recorded interviews with people are used almost as a counterpoint to the visual images of places or activities. This 'wildtrack' is often much more effective than synchronised sound, in which individuals are filmed giving an interview, not least because people become self-conscious in front of the camera. The lack of congruence between visual image and spoken comment is more subtle and more effective than a literal correspondence between the two. The power of personal documentaries is recognised by Denis Mitchell whose films *Night in the City* and *Morning in the Streets* remain as great artistic achievements and profound social comments about urban deprivation. He writes:

> I believe it is possible to express the essence of the human situation in our own time more effectively, because more truthfully, through the documentary than, say, through a fictitious drama . . . [what the producer manages to say] is in my opinion much more significant than what is said in a current affairs programme with a spoken, pseudo-objective comment.[7]

Why make a film about the Fens, a seemingly placid and peaceful agricultural region without any pressing social or environmental problems to catch attention? Our choice of region was determined by several considerations. We were anxious to move away from the familiar 'TV regions' – the Lake District and the Pennines – where audience interpretations would be coloured by their expectations about content. We needed a visually distinctive area, since the film itself would establish the location, and additionally a landscape sufficiently coherent so as not to confuse people with many changes in physical appearance. By no means insignificant was the fact that I grew up in the Fens and always had a strong intuitive 'feel' for the area, while the director, who did not know the region, found it a strange, exciting and compelling place. His commitment was crucial and, on the basis of his fascination, we began to explore the Fenland.

The initial preparation for the film involved extensive fieldwork by the director, Geoffrey Haydon, and myself. Norman Swallow, who directed several personal documentaries during the 1960s, commends intensive fieldwork to his contemporaries. Making *A Wedding on Saturday* took almost three months in the coal mining villages of South Yorkshire, for 'the producer/director, if he is to express the essential character and feelings of those whom he has chosen, must live among them for at least several weeks . . . to discover their true characters and their genuine (though often super-ficially concealed) attitudes'.[8] The academic might blanche at the bold pre-sumption that 'several weeks' is quite sufficient. In our case, the film (which took nearly three years to make) required many months of research, field-work, preparation and filming. The length of time reflected a desire to structure the film around seasonal changes in activity and a number of unexpected technical difficulties. The landscape proved inordinately difficult to photograph with the vast grey skies so characteristic of Fen country ending up as blank, white, featureless spaces on the film. Finding the right locations which capture the essential feel of the landscape on film was also problematic. Editing large quantities of filmed material to produce a coherent whole took the director and film editor many months more. Filming landscapes and activities was not carried out to any script or predetermined schedule, rather it was dictated by events. The chance discovery that a small farm was to be auctioned the next day meant bringing a lighting crew, cameramen and sound recordists from London at short notice. Similarly, great flexibility was needed to capture a spectacular skyscape or random event on film.

The Fens cover some 1300 square miles, stretching from just south of Lincoln to the edges of Cambridge, from Peterborough in the west to Lakenheath on the Suffolk borders. Within this large area, it is possible to distinguish the silt fen of Lincolnshire from the peaty area of the Black Fen (see Fig. 3.1). We decided to concentrate on the Black Fen, formed from drained swampy marshes and meres. Falling broadly within a triangle bounded by Peterborough, Wisbech and Cambridge, the Black Fen is the heart of the Old Isle of Ely administrative division. Even narrowing the focus to this area (some 730 square miles), we were aware of distinctive differences not so much in the landscape as in the culture of the inhabitants. The people of Lotting Fen around Ramsey do not consider that they have much in common with the populations of Burwell or Feltwell Fens around Ely. The Fenlanders who look towards Cambridge have aspirations and expectations that are different from those who turn to Peterborough.

For several months we were involved in intensive accidental fieldwork – a most exhilarating experience. A small number of initial contacts in the area led us to farmers, National Farmers' Union representatives, drainage board officials, publicans, retired Members of Parliament, Women's Institute members, auctioneers, poachers, roadmen, tenant farmers, factory hands, land workers, mystics, eccentrics and thieves. We met people who had never lived out of the Fens and people who had moved in from elsewhere, people

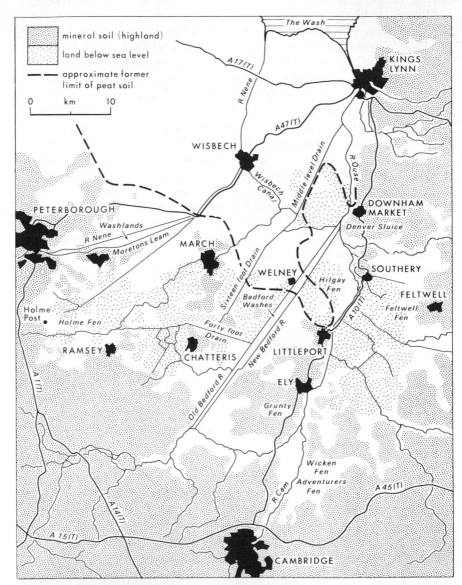

Figure 3.1 The Black Fens.

who had a great affinity for the place and people who wanted nothing so much as to leave. We talked to stoics, agronomists, conservationists, the apathetic and the involved, the hopeless and despairing, the angry and alarmed. We visited chip shops, local museums, factories, young farmers' meetings, pumping stations, farms, fields, dykes, droves, barn dances and discos, probation offices and gypsy encampments, strip clubs and gymnasiums, schools, churches, chapels and pubs. Almost without exception we were welcomed: people showed us their personal treasures and *memorabilia*, lent us books, photographs and home-made movies, allowed us to participate in family and farming life and agreed to be filmed going about their lives. Secondary sources of information lent support to much that we were told and substantiated many of our own impressions and interpretations. The meaning of the ordinary is indeed rarely obvious. The sources used included published recollections of life in the Fens; travelogues by locals and visitors at various dates; collections of folk tales and local myths; portraits of the Fens in literature; and academic works, notably those of H. C. Darby.

The sensory experiences which together combine to create a sense of place are usually lost in verbal accounts of character. The felicitous phrase or poetic insight may capture something of the feel of an area, but generally writings are unable to do justice to the quality of experience. Film allows one to recreate a living landscape through combinations of sights and sounds. 'A good documentary should tell not only what a place, or a thing, or a person looks like but it must also tell the audience what it would feel like to be an actual witness to the scene.'[9] Through the use of the camera and tape recorder, it is possible to take the audience with you through the landscape. A number of sequences in the Fen film were shot at water height moving along in a boat: reeds brushing the camera lens touch the shoulders of the audience; the startled call of a moorhen scurrying out from under the prow startles the viewer as well. The sense of immediacy and directness is retained in the translation of experience.

A number of themes began to emerge from our experiences with the Fens which seemed to encapsulate the sense of place felt by the inhabitants and visitors. Despite very many excursions up blind alleys, for example, filming Perkins' diesel factory, speedway racing, bell ringing, and the back streets of Wisbech, three remained constant – *leit-motifs* for the region. Of greatest significance, the Fens are a man-made landscape and the inhabitants are acutely conscious of their continuing struggle against the natural environment, particularly the permanent threat of inundation by flood waters. Secondly, the people who have made the place have a deep-rooted peasant mentality which is reflected in their overwhelming desire to own land, their exploitation of their natural resources, their aggression and cruelty. These characteristics are embodied in the myth of the 'Fen Tiger' (pp. 47–50). Finally, the strong feeling of isolation is an important key to the sense of place. The flat land separates rather than connects families, communities and villages. Traditionally, the area was cut off from the rest of England and this geographical isolation continues in the suspicious, hostile reception given to 'foreigners', i.e. people not raised in the Fens. These three themes were eventually to structure the film in the sense that the landscapes, activities and events were used as symbols for these interpretations of character.

A man-made landscape

More than any other landscape in England, that of the Fens represents the struggles and achievements of countless individuals against the elements. Their names have been lost but the landscape stands as their epitaph: 'Millions of faceless and nameless peasants and townsfolk the world over have through the centuries moulded, designed and designated their environments . . . having disappeared into the landscape much as their corporeal selves have turned into dust in some forgotten stretch of the Great Wall or ancient Fendyke.'[10] As so many Fen people said (not without pride), 'this is a man–made landscape, we have won our lands from the waters'. The present-day landscape is largely treeless with vast open fields stretching away in all directions under an immense sky – 'a similar land, repeating itself for ever.'[11] To stand on one of the small hills on the edge of the Fens in August is to look over a sea of corn and sugar-beet. It is a dramatic landscape with great scudding clouds, stunning sunsets and bitter winds. *'There is a saying round here that it's a lazy wind, it don't bother to go round you, it goes straight through.'* The distinctive sounds of the place include the wind, burning stubble, machinery working in the fields and skylarks. Most evocative of the sensory experience are the Fenland smells – burning stubble and bog oaks, reedy dykes and water.

Villages perch on small silt islands which stood out from the marshes, buildings lean at crazy angles due to the uncertain and shrinking peat. The villages and small towns like Chatteris and March are plain, functional and unadorned, refuges from the often bitter struggles down on the Fen. 'Whatever opinions we might hold of the unlovely clustered drabness of a Fen village, there comes easily to mind a sudden reaction on leaving it with a journey ahead in the gathering darkness of a winter night, as one enters the almost empty space of the Fens themselves. It is rather a feeling of going over the edge of beyond.'[12] Even the smallest change in height is crucial, since it provides safety from the flood waters known to local people as 'drownings', with the last great drowning occurring as recently as 1947. An obsession with altitude is reflected in conversation for people always talk of life 'down in the Fen' even though their own dwellings may only be five feet above sea level.

Flatness

Contemporary visitors to the area are struck most forcibly by the flatness in the same way that travellers in earlier centuries were appalled by the marshes and swamps. Flat land is not often considered beautiful; it is not valued by most people with preferences for hills, valleys and mountains. 'It is generally accepted that hilly, broken country from its very nature is beautiful and it is perhaps too little realised that there is a beauty of the plain alone . . . Light succeeds light in exquisite gradation, in endless perspective. The Sun's rays glinting in the middle distance stretch further than the eye can reach to the dim horizon, where far and misty, the plain ends.'[13] Vaughan Cornish, a geographer much concerned with the beauty of landscape, suggested that people find the Fens a stimulating scene and 'yet one which nobody would expect, least of all those brought up among the hills'. He described an encounter on a train to Ely with a Scottish gamekeeper who 'was as much thrilled by the sense of space as a newcomer is thrilled by the sense of height. "Whichever way I look" he said, "there is nothing to interfere with the view."'[14] In his memoirs of 40 years research in the Fens, the palaeoecologist Sir Harry Godwin describes the 'outstanding and, to some, entrancing quality of the Fenland landscape, an interminable flatness . . . [which] conveys feeling of vastness and remoteness'. Godwin believes that the flatness itself is the key to the affection for the area among its inhabitants and quotes a local who said that: 'any fool can appreciate mountain scenery but it takes a man of discernment to appreciate the Fens'.[15]

We found that local people were less effusive in their praise of the flatness, for the majority take it for granted and are not able to articulate what, if anything, they like about it or find beautiful. 'To be honest, I've never thought about it' was a typical reaction. Some people spoke of disliking other areas, frequently describing a feeling of being 'shut in' in hilly areas, others mentioned a sense of freedom in the Fens, and a few put forward the belief that local people have better eyesight because they can look further than anyone else! One lady said she could not stands hills − 'they block the view and you never know what funny people might come running down to carry you off', whilst another woman, whose windows look out over an expansive fen, said speculatively about the view 'well, perhaps I wouldn't mind a small hill to look at. Nothing too big though'. If the issue is pursued, many people will say that what they do like about the area is the peace, quiet and tranquillity of the Fens. The comments from one Fenman perhaps catch the key to the beauty for local people: 'The beauty comes from open, well drained, well cultivated land. There is a desire for straightness and accuracy, rows of plants. People feel in their bones that this is how it ought to be.'

Such insight gains support from Alan Bloom in his account of his struggle to bring the wilderness of Adventurer's Fen back into cultivation during World War II. He describes the intense satisfaction that he and his workers gained in burning, clearing and draining the Fen 'as men who were to be instrumental in changing the face and purpose of this locality, which had long been annoying and useless'.[16] He captures the beauty of Fenland farmed and goes some way to articulating the feeling of Fen farmers which others described as 'a land passion, a Fenland fixation'.

> I had prided myself on being used to the Fens, on having been born and bred where I could smell the tang the floods left on the grass . . . but I loved also the associations of fenland farmed; the green of winter wheat, the smell of black earth freshly turned, of twitch fires and the far-away smell of muck being carted from a dungle; the musical clattering of a mower, the rhythmic click-swish of a binder leaving the smell of wild mint behind; the sight of a long straight dyke and trimmed willow trees; and the wide horizons. All these things appealed, and were to be found in those fens where men had not ceased to struggle and to strive . . .[17]

A Fenland fixation

The desire for straightness and geometric precision in planting and tending crops is associated with intense cultivation. Every piece of available land is cultivated, trees and hedges have been grubbed out. *'Fenmen hate trees. They have to chop them down'* – ostensibly because they overshadow the crops, but in reality because they are a waste of space. Gardens are small and most often given over to vegetables, with perhaps a few token flowers. Deserted farmhouses and buildings down on the Fen are ruthlessly demolished to make way for more crops. People living on the edges of the Fenland interpret this desire to cultivate as indicative of the hard, greedy attitude of the Fenman which often repays itself: *'It's such valuable land, they cultivate it right up to the edges of the dykes. They are always falling off their tractors into the dykes and drowning themselves.'* The straight lines and lack of adornment are repellant for many outsiders. The somewhat jaundiced view of one Norfolk man reflects the opinion of many: *'I don't like its dykes; I don't like its straight roads. It's such dreary country especially in winter, it's such miserable countryside. You go out for a walk and what can you do, walk along the dykes – a mile out, a mile back, two miles out, two miles back? As far as you can go.'* He caught a similar feeling to that expressed by a London family who moved into a remote part of Hockwold Fen on the strength of cheap housing and a mistaken belief about the country life. They bemoaned the lack of footpaths and places to walk in pleasant surroundings: *'There is no landscape here. Only land.'*

Fen people will tell you that the lack of 'prettiness' is indicative of the hard lives they are forced to lead. Farming takes up every available moment of the day leaving no time to enjoy a country stroll. Their time is given over to the much more important task of growing food. As Sybil Marshall has said, in sanctimonious tones, 'The wild beauty may be gone, but it is replaced by a new beauty more useful to the over-populated world'.[18] However, the neglected appearances also reflect peasant mentality. Time and again we met ragged, scruffy-looking farmers who gave a realistic appearance of having no means of financial support, only to be told later *'Oh him, he's worth thousands'* – and most of it would be kept under the mattress. The wife of one farmer describes her life: *'Well, he comes home. Sleeps. Eats supper. Sleeps. We've never had a holiday because he can't be bothered and, like all farmers, he's mean.'*

One visitor to the area attributes the general lack of concern with appearances to the struggles of past generations: 'an inhospitable land where man snatched a difficult livelihood is an unlikely object for his beautifying care. Those who owned it did not live on it and those who acquired it did so with covetous hearts; . . . Untamed, it was unloved; and those who tamed it forgot to love it.'[19] The significant point about the feelings for land in the Fens is that it is not loved in the sense of creating pretty views or pleasant woodland glades or ornamental gardens; the land is valued for its usefulness, its productivity and its monetary value. The desire to own land is paramount in the minds of most contemporary Fenmen. We met many inhabitants who felt angry that people could no longer acquire smallholdings: *'You can't buy five acres any more, so you've got nowhere to start.'* As a farmer who owned 120 acres told us: *'There is a high proportion of agricultural workers who would dearly love to have a farm of their own. Don't stand a chance though. Farms are getting bigger and smallholdings are not available.'* The almost rapacious desire to own land creates a hostility between neighbours which is often barely disguised. Illness on the next farm is viewed with pleasure. It was said that: *'People round here don't ask how you are, they ask about your potatoes and hope they aren't doing very well.'* Many people now work in occupations removed from agriculture. The factory hands at Perkins diesel factory in Peterborough, for example, represent the first generation of Fenmen not to be tied to the land. Many are unhappy with factory life and keep smallholdings, others help out with the harvest and planting not least because *'I like to get my hands dirty. Keep in touch with the soil.'* The wife of a farmer who was forced to sell up because he was unable to find or afford labour to carry on was worried. *'He is too young to stop working'* she said *'and I don't know how he'll cope without the farm. It's in his blood.'* Her husband talked about his feeling about losing his land: *'It's my own land at the moment. When I look out of that window I own what I see. After Christmas I shall be trespassing on what was mine. I don't like it.'*

Water

The deep-seated fear among Fen farmers is the threat of flooding. The waters are kept at bay by an intricate system of drainage channels, dykes and sluices constructed over the past 300 years at great cost in terms of both lives and money. Much of the Fenland is several feet below sea level so water has to be pumped out to the Wash. Roads often run below the level of the water in the dykes creating a peculiar sense of anxiety: 'You stand above the landscape by five or six feet but so – you feel – does the impatient sea.'[20] The appreciation of the drainage network is acute among Fen people, for as one farmer announced *'These drains are our lifelines, all would be lost without them.'* Hilaire Belloc, however, viewed them in a less favourable manner: 'These dykes of the Fens are accursed things: they are the separation of friends and lovers . . . there comes another man from another part armed with public power, and digs between them a trench too wide to leap and too soft to ford. The Fens are full of such tragedies.'[21]

Pre-drainage, the Fens were a wild, watery and inhospitable place, with inhabitants living in mud and reed hovels on small islands and feeding themselves from the fish and wild fowl. The only surviving remnant of Fen landscape is in Wicken Fen, owned by the National Trust and kept as a nature reserve. This small area was the only way for the film to recapture the original appearance of the region. Historic accounts catch something of the desolate nature of the Fens and help to create a disparaging and hostile attitude among outsiders which lasted beyond the period of drainage. 'There is in many parts of our country, a vague idea that the Fens are really gloomy, unhealthy and swampy even at the present day [1878] . . . It is natural to infer that the description of a country abounding in muddy lands and black stagnant waters would convey the impression of insalubrity.'[22]

It is difficult to appreciate just how much the landscape had been changed. St Guthlac in the ninth century wrote of 'A fen of immense size . . . [with] now a black pool of water, now foul running streams and also many islands and reeds, and hillocks, and thickets' which was populated by devils with firey breath and tusks, conjured up from too much fasting. Daniel Defoe, coming to the Fens from Cambridge, wrote:

> As we descended westward, we saw the Fenn country on our right, almost all cover'd with water like a sea, The Michaelmas Rains having been very great that year, they had sent down great floods of water from the Upland Counties, and these Fenns, being, as may very properly be said, the Sink of no less than thirteen counties . . . In a word, all the water of the middle part of England, which does not run into the Thames or the Trent, comes down into these Fenns.[23]

The burden of being the sink of thirteen counties is not lost on contemporary

farmers who feel and express great resentment at having to pay for drainage
works, clearing out the dykes, and maintaining the pumps for 'highland
water'. *'The highland man pays no drainage rate and his water comes tumbling down
onto the Fens and we pump it out for him into the sea, and the Fenman has to stand the
cost of all this.'*

The history of the draining of the Fens is one of tremendous hardship for
the local people who were dispossessed of their common land and rights. Sir
William Dugdale, in his petition to King Charles II for more reclamation,
recognised the injustice but made light of it:

> As for the decay of fish and fowl, which hath been no small objection
> against this public work, there is not much likelihood thereof . . . so
> many great meres and lakes still continuing.[24]

The Adventurers who engaged in early reclamation, using the engineering
skills of the Dutchman Vermuyden, were rewarded with pieces of land while
the inhabitants were forced to leave. The drainage commenced in 1630 with
the cut of the Old Bedford River from Earith to Denver and continued
throughout the seventeenth century with the construction of the Twenty
Foot, Sixteen Foot and Forty Foot drains. The biggest task was the creation
of the Hundred Foot or New Bedford River, running parallel to the old,
which created a washland of some 5000 acres to act as a reservoir in times of
flood. The Middle Level drain was completed in 1848 and the last great mere
at Whittlesey was drained in 1851. The history of the drainage of the Fens is
that of improved technology, with windmills giving way to steam pumps
and eventually to electric equipment.[25] The achievement was recognised in
the piece of doggerel which rests on the Hundred Foot Engine:

> These FENS have oft times been by WATER drown'd,
> SCIENCE a remedy in WATER found.
> The power of STEAM she said shall be employ'd
> And the DESTROYER by itself destroy'd.

Or, in more constructed phrase, Darby[26] quotes a chronicler writing in 1685

> I sing Floods muzled, and the Ocean tam'd,
> Luxurious Rivers govern'd, and reclaim'd.
> Waters with Banks confin'd, as in a Gaol,
> Till kinder Sluces let them go on Bail;
> Streams curb'd with Dammes like Bridles, taught t'obey,
> And run as strait, as if they saw their way.

Most Fen people are aware that control of the waters is tenuous. The last
great Fenland flood in 1947 caused extensive damage. The breach in the bank
at Southery was remembered as *'the sound of an explosion and then the water
rushed in, just like a steam engine.'* This flood could be considered 'an Act of God'

but in earlier times the Fen would have been flooded intentionally by land-owners concerned to protect their own lands. Many of the folk tales in the Fen record faithfully the condition of life and preoccupations with flood-waters. Chaser Legge, in his stories of the 1861 flood, tells how Fen people would mount bank watches to prevent the gentry from blowing up the banks and flooding the peasants in their hovels down on the Fens. As he says: 'ever since the Fens were drained, whenever the Little Ouse topped up, someone would come down and blow the banks. They reckoned this was a poor old fen anyway and if it got flooded then the ones higher up would be safe. My grandfather used to say "don't trust anyone when the river is full"'.[27] Such distrust is deep in the folk memory of Fenmen and still comes out in conversation, for floods are an ever present threat. A smallholder who gave up farming said: *'If I had my time again I would have done it different. You didn't buy land then because there wasn't any money and you were frightened of debt and you got into debt because you were flooded virtually every year. I had £400 once and said, right we'll put that on one side but the wife said put it into potatoes. But I shouldn't have listened to her, it was a wet year, we got flooded. So that was gone. As if we had never had it.'*

The supreme irony of the Fens is that the peat land, won at such cost and effort, is disappearing. Drainage has resulted in the fall of the water table and as the peat dries it both shrinks and blows away. A particularly distressing phenomenon of spring time is the peat 'blow' when great dust storms sweep across the Fens. *'It's like black pepper. It just seems to seep under closed doors and windows and covers everything.'* The blows also damage young crops and remove seed. Peat shrinkage is held up as an indication of the rapaciousness of the farmers by non-agricultural people in the area, and the Post in Holme Fen, which now stands some 13 feet above ground level, is a symbol of their greed. For farmers anxious to hand on their land to their children, the thought of watching their lifeblood blow away and shrink is too horrifying to contemplate. *'It frightens us to think about it sometimes.'* Yet there is little attempt at soil conservation partly because of a belief that it is beyond their power. *'Shrinkage'* we were told with great authority *'happens because it's all to do with bacteria eating the peat.'* Peat is a difficult substance to handle, being friable and readily combustible. *'I'd not had much experience with peat farming; I lit the straw and by a cruel trick of fate really, the wind changed and it swept across the whole farm and before long there was sixty acres of turf burning.'* This farmer's neighbour told us the same tale with much amusement at his misfortune. Farmers are often reminded of the antecedents of their land by bog oaks – great trunks of trees, often 100 feet long, which lie buried in the peat. Machinery is broken and much time wasted while these remnants from a former age have to be dug out, carted away and burned. The stoical attitude towards environmental hazards is interpreted by outsiders as apathy. *'These locals see themselves as victims. Because of the seasons, they have to stand back and let it happen.'*

The Fen Tiger?

A strong notion of environmental determinism is evident in interpretations of Fen character. Both local inhabitants and outsiders attribute many of the past and present characteristics of the people to their environmental conditions. 'Fen folk can be coarse, dour, close, clannish and untrusting, having little respect for persons or niceties, lacking an appreciation for art and beauty. Many are materialistic to a degree but they are also realists.'[28] It is often suggested that these characteristics are inherited through the struggles between locals, and between people and the forces of nature. The conditions of life horrified many visitors in the times before drainage and well into the twentieth century. Many people still describe the old Fenmen in terms of the environmental hardships they had to bear. *'Look at your typical Fenman, he was a poor little bent old man, full of rheumatism. The damp, it must have affected him.'* W. H. Barrett was anxious to place on record the stories of men who were the last of the real old Fenmen, to give outsiders some idea of what life was really like in the desolate Fenland a century ago. The lack of culture is evident still as we found virtually no interest in music or art; indeed Fenmen rarely express themselves in song except when filled with tavern ale or religious fervour. The heritage of present day Fen people seems close to those of their ancestors who were 'a savage race by the end of the seventeenth century, living in isolation and harassed by disease and floods.'[29]

Fen people were isolated from the rest of the country by the dreadful physical conditions. Dugdale perhaps overstates the hardships experienced by people, for he was anxious to press his petition for drainage:

> In winter the inhabitants upon the hards and banks within the Fenns, can have no help of food, no comfort for body or soul; no woman aid in her travail, no means to baptise a child, or partake of the communion . . . and what expectations of health could there be to bodies of men where there was no element good, the air being for the most part cloudy, gross and full of rotten harrs; the water polluted and muddy, yea full of loathsome vermin, the earth spongy and boggy and the fire noiseome by the stink of smoking hassocks.[30]

Daniel Defoe also felt great concern for the Fenland inhabitants and was not a little surprised at their stamina: 'One could hardly see from the hills and not pity the many thousands of families that were bound to or confin'd in these foggs and had no other breath to draw than what be mixed with these vapours and that steam which so universally overspread the country . . . [yet] those that are used to it live unconcerned and as healthy as other folks except now and then an Ague.'[31] Ague was indeed the blight of Fenmen, resembling a form of arthritis or rheumatism, in which they were taken with fits of shaking which, it was commonly believed, could be cured by a strong dose of gin or brandy. Charles Kingsley was appalled at the intemperate habits: 'The foul exhalations of autumn called up fever and ague, crippling and enervating, and tempting, almost compelling, to that wild and desperate drinking.' He extolled the benefits of drainage: 'at least we shall have wheat and mutton instead, and no more typhus and ague; and it is to be hoped, no more brandy-drinking and opium-eating.'[32] Opium, the consolation of the poor well into the 1920s, was considered to be a vice peculiar to the Fens. In the present day, newcomers express surprise at the amount of under-age drinking and general drunkenness they see around them: *'But it's oblivion and that's why the kids start drinking at fifteen.'* Many feel that people seek oblivion to get away from the drudgery of life: *'You see kids of twelve driving tractors. They are forced onto the land. Every waking hour they're working.'*

The boredom of youth and lack of opportunities in the area are characteristic of many isolated rural areas. Fashions are behind the times, trends are missing. One group of lads who formed a pop group played in Cambridge to a constant barracking from the students: *'They treat us as if we are a bunch of yokels.'* Young people want to get away. Crime is a major concern for local people, as is the drug problem in Wisbech. Often exaggerated in newspaper reports, the problems are put into perspective by a March probation officer: *'Well, there is hard organised crime in Wisbech. March has a few inexperienced crooks that everyone knows. And Chatteris has just twelve hooligans.'* Many people, however, speculate about the possible causes: *'Well I think it's all to do with the depressing countryside. They get bored.'* The lack of employment opportunities in the area also disturbs people. March refused to allow the Metal Box Company to build a factory in the town, and Wisbech has not encouraged light industries which now go to Kings Lynn or Peterborough in preference. *'I think they have been incredibly stupid, they have lost out on jobs and now they are losing their young people.'* Certainly the bright children leave the area. Those who remain can choose seasonal land work, but the farmers prefer women. It is arduous and back-breaking work, often on hands and knees. Work is available in pre-packing and food-processing plants where vegetables are graded, but it is dirty, unrewarding work which is poorly paid and which offers no future. Many times, we met an apathetic air of resignation from those left behind.

There is considerable social distance between inhabitants and newcomers. Some people are prejudiced against 'foreigners' (see above): *'I don't mind outsiders, but not too many of them.'* The general response to outsiders is to ignore them. *'These outsiders, people from London, think local people are stupid country bumpkins. So local people tend to play up to them and get better deals. In the main, they might just as well not be here.'* This comment supports exactly the experience of a Californian living on the edges of Wisbech who said that as an outsider, he might just as well be invisible. Newcomers find the Fen people unfriendly and anti-social. *'Life here is bleeding terrible'* confessed one Londoner who bitterly regretted his decision to retire to March. *'They are*

*anti-social. They never go anywhere. Think if they go to Kings Lynn they'll fall off
the edge of the world.'* The common experience is one of suspicious and shifty
locals who will 'do you' if they can. Fen people are described as hard, greedy
and lacking in sentiment, like the farmer who, when his gun dog failed to
retrieve a bird, shot the dog then and there.

Fen people will acknowledge that there is a streak of cruelty and a lack of
compassion in their nature, but a semi-retired poacher justified it thus: *'We
had to eat anything then – birds, eels, bream, pike, moorhen, sparrows, anything that
was alive. We had to eat them or starve because there weren't no money about. We
were cruel, it's no good saying we weren't. Ain't got no sympathy for nothing in them
days, not even for ourselves.'* He was talking about life in the early decades of
this century, but the Fenman's reputation for cruelty and savagery has a long
tradition which is encapsulated in the myth of 'Fen Tiger'. *'The old Fenmen,
what we call the old Fen Tigers, I mean, they were a tough breed. They were isolated
communities, at war with each other, always fighting.'* We failed to find anybody
who claimed to be a Fen Tiger and most people seemed to be embarrassed to
be asked. Well, what were they like? 'I don't know why the old Fenmen were
allus called Tigers, unless it were because they used to act so wild and shy, not
being used to seeing many folks or whether strangers thought they looked a
bit fierce. There is a saying you can use about any man who got a good crop
'air (specially if it's on his chest) and a good set o'teeth. You can say "'E's all
'air and teeth like a Ramsey man".'[33]

At least part of the myth stems from the fierce reactions of Fenmen to the
activities of the reclaimers and those in the pay of the landowning aristocracy
and the church. The peasants living down on the Fens would cheerfully
attack the representatives of the authorities. 'Big farmers daren't live there;
they put foremen in to run the farms and the farmers only came into the fen in
daylight and they rode back to the towns, where they lived, before dark, and
ganged up for the ride home with others like themselves . . . The parsons
were as bad, if they weren't worse. They wanted to keep in with the gentry so
they used to tell the people to put up with their miseries and not grumble,
then, when they got up top, they'd be ever so happy listening to the sound of
harps.'[34] The established church had little compassion or concern for the
poor. As a result of the Littleport riot in 1816, five Fenmen were hung. Their
bodies lie in St Mary's Church in Ely under the plaque: 'May their awful fate
be a warning to others.' It seems such a short time ago: in 1979 a legal battle
was fought in Ely between two descendants of the main protagonists over an
attempt to get the gravestone moved back to Littleport.

Feelings about the established church run just below the surface and Ely
Cathedral is for many a symbol of exploitation. The chapels were the source
of the true religion in the Fens, a fact acknowledged by one vicar we spoke to
who agreed wholeheartedly with one of his forebears: *'A minister condemned to
live in such a place must be a man of iron nerve or unrefined taste.'* A wave of
religious fervour swept the Fens in 1900 when many people were baptised in

the rivers. 'The converts were led into the water till it was up to their waists, then ducked under and they scrambled out as the believers sang hymns and the unbelievers expressed their opinion on those being ducked, stating it was the first good wash they'd had for months.'[35] There is a contemporary revival of Baptist fervour in the Fens near Cambridge. The preacher would like to reinstate the habit of ducking in the river but is unable to, because the water authority keep the level of the sluice too low! How far the revival is a resurgence of old beliefs and how far a response by the young to the charisma of the preacher is difficult to judge.

Isolation

Fen people may indeed prefer a personal God but they also feel isolated from the established church and its services. Indeed, the isolation is a major experience of life in the Fens, and it is no small irony that the flat land separates rather than connects communities. Many of the houses and recent buildings which straggle along the high roads reflect an unwillingness among the women to put up with life down on Fens.[36] Typically, the farmhouses were connected with the main road by a soft, peaty drive which became impassable in winter. Women and children would be marooned in the Fen, sometimes for many months. One teacher spoke of the *'special isolation of women and children. Most of the children down on the fen are semi-literate because they never hear any conversation.'* Many people acknowledge that they left school unable to read and write because they missed so much schooling. One describes the children as having *'a quiet reticence'*. The isolation has led to a recognisable and sizeable problem of mental illness in the Fens. Many people will quite happily admit to being on ten or fifteen milligrams of valium a day 'for depression' – and drugs seem to have replaced a reliance on opium and poppy head tea.

The traditional isolation of farmsteads in the Fens encouraged incest. Generations of inbreeding have promoted the growth of various genetic diseases such as Huntingdon's Chorea: sufferers tend to be somewhat retarded. March has recently opened a hostel for 25 mentally subnormal people. In previous generations, these people would have been given menial tasks on the land and cared for within the family. The mechanisation of farming has greatly reduced the demand for unskilled labour and brought to light this previously hidden problem. However, people have long been aware of the apparent simple-mindedness of Fen people, 'slow in speech, they were often thought by strangers to be dull witted' and one Fenland physician attributed the subnormality to poppyhead tea: 'In frequent use and taken as a remedy for the ague . . . to children during the teething period poppyhead tea was often given and I do think this was the cause of the feeble-minded and idiotic people frequently met with in the Fens.'[37] Apart from genetic retardation, there is a recognisable psychological illness within the Fenland community which many people believe is caused by the isolation experienced down on the Fen and the flatness: people just give up. A Cambridge psychiatrist describes the symptoms: 'We call it cultural retardation. People out there are often retarded, but it's not so much an illness as a way of life; you only find out about it when they are struggling with our city ways in Cambridge. Then the pressures and expectations of the urban existence are too much for the fenman and he breaks down. In his own context he can operate quite successfully . . . The "fen syndrome" is a way of coping with the isolation.'[38] A recent migrant to the Fens captured well the sense of isolation and desolation experienced by many in the area when he said: '*There is so much sky. It has an effect on the psyche. You are all the time waiting for something to happen.*'

The sense of place

Looking back now at the experience of researching for the film, I realise that we nearly came to grief in the Fens. The involvement with so many people, combined with the sheer volume of information we gathered, and the increasing complexity as we became more deeply involved with the social undercurrents of the area, threatened to overwhelm the film. It was essential that the story be crisp, clear and relatively straightforward for there would be no narrator to point the way; the film was only to be 50 minutes long and, perhaps most important, it was to entertain people as well as telling them about life in the Fens. Many aspects of the economic and social conditions which would be essential to any sound geographical interpretation would complicate the story and confuse the audience. My involvement with the film ended with the identification of those themes which provided the key to the character of the Fens. It was the task of the director and film editor to take all the information on film and tape-recorded interviews and re-create the sense of place. 'It is not enough to show bits of truth on the screen; separate frames of truth. These frames must be thematically organised so that the whole is also a truth.'[39]

As befits the title, the film is primarily a sensual experience. The atmosphere of the Fenland is caught in beautifully photographed landscapes and the commentary is constructed from the sounds of the Fens – birds, reeds, the wind, stubble burning, pumps working, plainsong in the cathedral and the sounds of machinery – interspersed with occasional reflective comments from Fen people themselves. Two threads running through the film provide the basic structure of the narrative. The first is the experience of the physical environment, its visual qualities and characteristic sounds. There are no villages or towns in the film, for the essential Fen remains the landscape. The second thread is the agricultural life of the area, and seasonal changes, starting in late summer and ending in late spring, provide the main story. Crops are harvested and processed, land is ploughed and sown, dykes are cleaned, and plants and animals tended. Within this framework, scenes are used as symbols to express the interpretations of character by the inhabitants and the beliefs of the director. Geoffrey Haydon experienced a strong feeling of empathy for the young and the poor in the Fens who, he felt, were trapped by the sheer drudgery of agricultural life and who were unable to break away. The heart of the film, at the end of the winter sequence, shows a youth shovelling carrots in the mud under a dripping conveyor belt. The scene has the sound of a factory running beneath it; it changes to the interior of a canning factory with sheets of tin flowing like water through the cogs of machinery and then, finally, cuts to a churchyard with a fallen cross and gravestone.

Objects and events are used as symbols of Fen character – the teeth of a combine, the spikes of an eel glave, tearing flesh from a rabbit and spraying water lilies in the dykes, dead moles hanging on barbed wire and eels trapped in a box convey ideas of cruelty, utility and exploitation. Impoverished-looking farmers pick through the detritus of a farm sale, stubble burns, and bog oaks are hauled from the ground and fired. All these activities happened in the remote landscape. The flat lands and enormous skies are photographed with telegraph poles marching into the distance. Endless dykes flow through the picture; men working in fields are engulfed by the sky and deserted houses lean at peculiar angles in vast, empty fields. The Fenland obsession with water grows through the film. The camera moves along dykes and drains, records a baptism with the total immersion of the convert; observes men clearing and cleaning the drainage channels, and watches sluices emptying, pumps being primed and washlands flooding. Archive film recaptures the horror of the 1947 flood and the film ends, as it begins, with the indistinct boundary between the land and sea in the Wash.

It has been argued recently that the familiarity of television and its significance in the development of popular culture 'makes it so important, so fascinating and so difficult to analyse'.[40] In this essay, I have commented on my experiences in the Fens; the final question must be to what extent these interpretations reflect the experiences of the inhabitants. A number of films made for television have deeply offended the inhabitants of the places depicted in them. Denis Mitchell's film about Chicago, for example, was banned in that city for six years because it was considered dishonest, distorted and disgusting by local representatives. A Wedding on Saturday, mentioned earlier, led to complaints from local councillors because it neglected to show 'such splendid amenities as parks, swimming pools, maternity hospitals and the like'.[41] A drama documentary, The Land of Green Ginger, which was about the life of fishermen in Hull, created such intense anger among local people that a public meeting was organised to take Alan Plater to task for having written it. In all three cases, the films failed to emphasise the environmental qualities and amenities valued by the inhabitants. The television companies are blamed for first creating and then perpetuating a bad or unfavourable image of the place.

The beliefs and impressions that outsiders have of places – their evaluations of the beauty or ugliness of the environment and assumptions about the character of its inhabitants – are undoubtedly influenced by the style and content of media reports. The responsibility for the selection of information and the way in which that information is moulded into a programme rests with the journalist or film director. It may well be that the message is essentially truthful, although not what the local people want to hear. *A Sense of Place* can be criticised for being too selective in that it did not include any material about life in the towns and villages. It may also be criticised for having insufficient commentary and not making its exposition more clear. I would suspect that some viewers may have found it boring, but I do believe that it is a faithful reflection of the interpretations of regional character and values that the inhabitants themselves have of the Fens.

Notes

1 *A Sense of Place: The Fens* directed by Geoffrey Haydon, produced by David Collison, BBC 1980. Other regional films made by the BBC include three films with John Betjeman, entitled *Scotland – The Lion's Share; Wales – More than a Rugby Team; A Prospect of England*, 1977. *The Making of the English Landscape* twelve films: Series I: *Ancient Dorset; Conquest of the Mountains; Marsh and Sea; Landscape of Peace and War; The Deserted Midlands; The Black Country.* Series II: *Behind the Scenery; The Fox and the Covert; No Stone Unturned; Brecklands and Broads; The Frontier; Haunts of Ancient Devon*, 1978. From Independent Television, *A Writer's Notebook: The Pennines*, four films with Ray Gosling (1979).

2 Meinig, D. W. 1979. Reading the landscape: an appreciation of W. G. Hoskins and J. B. Jackson. In *The Interpretation of Ordinary Landscapes*, D. W. Meinig, (ed.), 236. New York: Oxford University Press.

3 Smith, A. 1973. *The Shadow in the Cave*, 6. London: George Allen and Unwin.

4 Swallow, N. 1965. *Factual Television*, 176. London: Focal Press.

5 ibid., 180.

6 Grierson, J. quoted in Sussex, E. 1975. *The Rise and Fall of the British Documentary*, 206. Berkeley: University of California Press.

7 Mitchell, D. quoted in Swallow, N. op cit., 177–8.

8 Swallow, N. op cit., 185.

9 Stryker, R. quoted in Stott, W. 1973. *Documentary Expression in Thirties America*, 29. New York: Oxford University Press.

10 Samuels, M. S. 1979. The biography of landscape. In Meinig, D. W. op. cit., 81.

11 Belloc, H. 1906. *Hills and the Sea*, 95. London: Methuen.

12 Bloom, A. 1953. *The Fens*, 29. London: Robert Hale.

13 Wedgewood, I. V. 1936. *Fenland Rivers: Impressions of the Fen Counties*, ix–x. London: Rich and Cowan. See also Shoard, M. this volume.

14 Cornish, V. 1943. *The Beauties of Scenery*, 48. London: Muller.

15 Godwin, H. 1978. *Fenland: Its Ancient Past and Uncertain Future*, 1. Cambridge: Cambridge University Press.

16 Bloom, A. 1944. *The Farm in the Fen,* 96. London: Faber and Faber.
17 ibid., 19.
18 Marshall, S. 1977. Fen Tiger. *Vole* **3,** 46.
19 Wedgewood, I. V. op. cit., 45.
20 North, R. 1977. Cycling without hills. *Vole* **3,** 47.
21 Belloc, H. op. cit., 94.
22 Miller, S. H. and S. B. J. Skertchly 1878. *The Fenland: Past and Present,* 413–14. Wisbech.
23 Defoe, D. 1714. *A Tour thro' the Whole Island of Great Britain, Vol. 1. Eastern England,* 119–20.
24 Dugdale, W. 1622. *The History of Embanking and Draining of Divers Fens and Marshes Both in Foreign Parts and this Kingdom.*
25 Hills, R. L. 1967. *Machines, Mills and Uncountable Costly Necessities: a short history of the drainage of the fens.* Norwich: Goose & Sons.
    Mason, H. J. 1973. *An Introduction to the Black Fens.* Ely: Mason.
    Astbury, A. K. 1958. *The Black Fens.* Cambridge: Golden Head Press.
26 Darby, H. C. 1956. *The Draining of the Fens.* Cambridge: Cambridge University Press.
27 Barrett, W. H. 1963. *Tales from the Fens.* London: Routledge and Kegan Paul.
    See also Porter, E. 1969. *Cambridgeshire Customs and Folklore.* London: Routledge and Kegan Paul.
28 Bloom, A. op. cit., 148.
29 Barrett, W. H. op. cit., ix–xiii.
30 Dugdale, W. op. cit.
31 Defoe, D. op. cit., 121.
32 Charles Kingsley, quoted in Miller and Skertchly op. cit., 421.
33 Marshall, S. 1967. *Fenland Chronicle,* 8–9. Cambridge: Cambridge University Press.
34 Barrett, W. M. op. cit. 87–8.
35 Barrett, W. H. 1965. *A Fenman's Story.* London: Routledge and Kegan Paul.
    For other reminiscences, see Randell, A. 1969. *Fenland Memories.* London: Routledge and Kegan Paul.
36 See also Chamberlain, M. 1975. *Fenwomen: A Portrait of Women in an English Village.* London: Virago.
37 Lucas, C. 1930. *The Fenman's World: Memories of a Fenland Physician.* Norwich: Jarrold.
38 Garvey, A. 1977. The Fen Tigers. *New Society* **48,** 429–30.
39 Vertou, D. quoted in Barnouw, E. 1974. *Documentary: A History of the Non-fictional Film,* 57. New York: Oxford University Press.
40 Fiske, J. and J. Hartley 1978. *Reading Television,* 6. London: Methuen.
41 Swallow, W. op. cit., 185.

4 *The lure of the moors*

MARION SHOARD

Heather and grass moorland covers a third of the land surface of Britain: it is not in short supply. Nor is it unique to Britain: much vaster expanses of moorland blanket tracts of Scandinavia, Russia and Canada. Yet in Britain, the moors have captured a unique place in the imagination of many members of the countryside establishment. So much so, that the idea of protecting moorland has dominated countryside policy-making for almost the whole of the post-war period – at the expense of other types of landscape whose need has been greater. All ten of our national parks, for instance, have been selected to enshrine moorland, even though the official criteria for park designation suggest that remote moorland is far from the ideal candidate for national park status. While these parks were being designated, Britain's traditional lowland countryside – the patchwork quilt of fields, woods, downs and marshlands, separated by hedgerows, banks and winding streams – was undergoing a mounting onslaught from agricultural change. This lowland countryside is England's most distinctive landscape type, and survey evidence suggests it is the type most popular with the general public.

Yet in 1979, even after 30 years which had seen an agricultural revolution wipe out the character and recreation potential of huge tracts of lowland England, the Labour government's aborted Countryside Bill included tighter restrictions on the ploughing of moorland in national parks as its only real new proposals for landscape conservation. The Conservative government's 1980 Wildlife and Countryside Bill also confined itself largely to moorland in national parks as far as landscape protection was concerned. It was only in the debates on the Bill and amendments tabled to it that the needs of other forms of landscape began to be considered seriously. Today, any attempt to rethink countryside policy-making is still bedevilled by the long-standing presumption that the moors must take priority. So how has moorland managed to exert the spell it has over Britain's rulers for a quarter of a century? It is a long story and its roots lie in the history of landscape conservation in Britain and the psychology of our conservationist classes.

Tom Stephenson, who is now 88 years old, has devoted his life to preserving wild moor and mountain and opening it up to the public. In an article in the now defunct *Daily Herald* in 1935, he called for a public footpath running the whole length of the Pennines. Fourteen years later, Parliament lumbered into action, and today the Pennine Way is one of eight long-distance paths which have opened up 1500 miles of the English and Welsh countryside to walkers. In 1952, Tom Stephenson became the full-time secretary of the Ramblers' Association; he held the post for 21 years, retiring at the age of 76. He is still an active and influential member of conservation and recreation pressure groups.

For Tom Stephenson, it all started one clear, frosty, winter's day in early March 1906, when he was just 13 years old. He was living in a small Lancashire town called Whalley, lying in the Ribble Valley and hemmed in on either side by two great moorland massifs – the Forest of Pendle and the Forest of Bowland. On that bright March day, young Tom Stephenson climbed 1830 feet to the summit of Pendle Hill. Beneath him to the south he could see a great range of factories with chimneys belching out smoke that blanketed towns like Nelson, Colne and Burnley. The other way, as he put it to me in an interview in 1978: '*It was just wild country, nothing at all. And the great attraction was that so easily you lost any sense of industrialisation or civilisation; you felt you were alone in the world.*'

In the same year (1906) at the opposite end of England, a two-year-old girl, Sylvia Pleadwell, later to become Lady Sylvia Sayer, was making her first attempt to ride a Dartmoor pony. Unlike Tom Stephenson, Lady Sayer did not undergo a sudden conversion to wild country. She spent her childhood moving from one house to another in Plymouth, Portsmouth and Greenwich, but her grandparents had a house in the middle of Dartmoor. She spent holidays at Huccaby House from her earliest years, and Huccaby, set in wild moorland and with a Dartmoor stream within earshot of her bedroom, was the constant and magical place to which she longed to return.

Now 77 years old, Lady Sayer is writing her memoirs. Entitled *Granite in my Blood,* they describe a lifetime's devotion to Dartmoor and battles to protect it from mining companies, water authorities, farmers, the Army and local councils – most of them fought during the 22 years she was chairman of the Dartmoor Preservation Association. She told me what she remembered of her earliest trips to Dartmoor: '*I was born in Plymouth and . . . one did see in those days fields outside Plymouth and a little village where we were sometimes taken for church. But none of that meant anything. It was the moors, it was Huccaby . . . Plymouth in those days was rather a smoky town. One never noticed that until one stepped out of the rather tiny train at Princetown station. As a child the very air was magic because it was so different, so clean and pure and absolutely heady . . . But it was of course primarily the wildness, the feeling of freedom . . . It was the freedom of it all which was so wonderful . . . Instead of being in a terraced house in Plymouth there was all this lovely wild freedom.*'

Kate Ashbrook is 24 years old. Born and brought up in Denham on the edge of London, she fell in love with Dartmoor at the age of twelve. This is how it happened: '*It was a summer's evening and it was sunny but also hazy, and I remember being up on Hameldon, which is this great long ridge, and looking out over Dartmoor you couldn't see anything very clearly, just outlines of the hills. We were having a picnic supper and they said have a race to Hameldon Beacon but I just didn't feel like racing, I wanted to walk on my own. I felt how wonderful it was that here's Dartmoor, such a place existed and here I was experiencing it and what a wonderful thing it was. It made me feel very happy but emotional too: I remember crying.'*

Kate chose Exeter as her university solely because of its closeness to Dartmoor. Since the beginning of her undergraduate days, she has devoted all her spare time to working for the Dartmoor Preservation Association. Sylvia Sayer recognises in Kate the same love of Dartmoor that inspired her own lifetime's devotion to its preservation. Kate loves to ramble over Dartmoor, summer and winter alike. What does she think about out there? *'I have a feeling of freedom on Dartmoor. Because it's far from civilisation. It does bring you back in proportion. You've been fussing over some beastly something or other, like exams, and it does make you think there's a lot more to life than your job. It has a wonderful therapeutic effect.'*

I talked at length to five lovers of moorland, of different ages and from different walks of life, to try and find out what it is about the moors that inspires such fierce devotion. Tom Stephenson, Sylvia Sayer and Kate Ashbrook I have already described. The other two are Gerald McGuire and Malcolm MacEwen. Gerald McGuire was born in London in 1918. He served on the Countryside Commission from 1976 until 1980 and has worked for most of his life as an officer of the Youth Hostels' Association. He served on the North York Moors National Park Committee for 19 years and has been a leading figure in the Council for the Protection of Rural England for about the last ten years. Malcolm MacEwen was born in Inverness in 1911. He is a newcomer to the countryside movement and, unlike the others I talked to, he campaigns mainly alone. A journalist, broadcaster and architect by trade, MacEwen was appointed to the Exmoor National Park Committee in 1973. Since then he has fought hard to halt the ploughing up of the heather moorland of Exmoor, and it was his efforts more than those of anybody else that led the government to set up an inquiry into land use on Exmoor which reported in 1976, calling for greater control over moorland 'reclamation' in the national park.[1]

What has led these five people to devote themselves to protecting what many of their fellow citizens see as bleak, dull, forbidding wastelands? It is clear that all five see the moors in the same way. They regard them as a refuge whose remoteness cuts them off from the man-made environment in which they spend most of their lives. They find on the moors, and some of them on mountains too, what I shall call 'wilderness'.

'Wilderness' is not always just what it might seem. Some people can feel quite apart from the rest of us merely by wandering on a well-vegetated piece of derelict land in London; others need to get hundreds of miles away from the nearest town. For the people I talked to, the only environments that provide a spiritual cocoon strong enough to keep out Man's works are moorland and, for some of them, also mountain. Because they believe that the experience of 'wilderness' is central to human well-being, they become almost fanatical in promoting the conservation of what they believe to be our most important landscape types.

In the United States, where wilderness preservation completely dominates countryside protection policies, wildernesses are primaeval, usually forest landscapes.[2] Not so in this country. People calling for the preservation of wilderness in this country are not seeking to preserve our oldest landscapes – those remains of the original post-Ice Age forest cover that still exist mainly in the Weald, Devon, Essex and Suffolk and are about 12000 years old. Nor are they trying to preserve the first hedgerows with which our forefathers enclosed land. They are seeking to protect moor and mountain, although most moorlands are relatively recent landscapes, created at most 4000 years ago through the destruction of forest to provide wood, charcoal or sheep runs. What is more, most moors rely on Man's activities – burning and the grazing of his animals – for their continued existence: left to itself, heather moorland reverts to scrub or woodland within about 60 years.

Ecologically, moorland is not a rich habitat; some ecologists have gone so far as to label moors 'biological deserts', compared with habitats like deciduous woodland and chalk grassland which contain a far greater variety of plant species per square metre. The reason for the lack of species variety of the moors is the way in which they are managed: grazing and burning operate selectively on the vegetation causing an increase in the number of plants resistant to these processes – heather, brackens, the moor grasses and fescues – at the expense of plants that cannot survive the continual removal of their leaves and stems, such as tree seedlings. On large parts of Bodmin Moor, heavy grazing and repeated burning over thousands of years have now destroyed even the heather, leaving only a few species of tough grass.

Pedigree, then, is not what makes 'wilderness' in this country at least; so what is it that defines 'wilderness' for the band of people in England and Wales who seem to associate the idea with moor and mountain? Unlike the gregarious bee or the solitary wolf, Man is both a herd animal and a loner. Since the dawn of the Romantic era, however, it is the human individual who has been venerated in the West, and we have all felt obligated to seek ourselves. Those attracted to wilderness landscape seem to be seeking a context for the pursuit of their individual identity away from the herd. To do this, they need to get away from the environment their fellow men have created for the group to a place as devoid as possible of what is obviously human handiwork. (It apparently does not matter if the landscape is in fact man-made – like a grouse moor – so long as it looks 'natural'.) A variety of other living things is also unhelpful: what is sought is a blank canvas on which individuals can commune with themselves or their Maker.

Primary conditions for wilderness

There seem to be seven conditions, all of which landscape *must* meet if it is to arouse the enthusiasm of the wilderness lobby in this country. These conditions are: wildness, openness, asymmetry, homogeneity, height, freedom for the rambler to wander at will, and the absence of what is obviously human handiwork.

Wildness. Wildness, the antithesis of domestication, is the key quality of moorland in the eyes of its admirers – such as Tom Stephenson, who was attracted by the wildness of the Pennines when he first saw them from Pendle Hill 75 years ago. He has little time for the neat, tidy farmed landscape of lowland England: *'I've learnt to endure lowland scenery. I realise that you can't help but admire a mountain, but to appreciate the more subtle lines of lowland landscape is more difficult – I think it's got to be acquired.'*

Tom Stephenson now lives in Buckinghamshire close to the Chiltern hills. He often drives up to them for a walk, although their scenery does not attract him; the Chilterns are too tame for his taste and he visits them only for the exercise. For Tom Stephenson, the wilder and more rugged the country the better. Above all landscapes, he loves the Cuillins of Skye – moorland out of which rise rocky crags – because *'they are gaunt, almost black at times – gabbro and basalt. They are shattered and pinnacled and you can see almost from one side of the mountain to the other through cracks in the rock.'*

It is the wilderness of moorland that writers like Emily Brontë and Sir Arthur Conan Doyle, who have set novels in moorland, emphasise: their creatures of the moors – Heathcliff in *Wuthering Heights* and the eponymous hound in *The Hound of the Baskervilles* – are wild, tempestuous spirits, whose temperaments reflect the environment that nurtured them.

Openness. All five of my interviewees consider the 'openness' of moorland central to its appeal. For Gerald McGuire at least, the fact that moorland landscapes are more open than mountain ones means that they are better. *'The appeal of moorland in contrast particularly to mountains is this openness: this great vista and you're in the middle of it and you're preferably very much alone in it. Associated with the openness is the sky. You've got your great vista of moorland and you've got what seems a big sky. Northumberland is the country where this is somehow even more marked than North Yorkshire: these great wide skies.'*

Gerald McGuire now lives in Hemel Hempstead in Hertfordshire, and after work in the evening he chooses to go for a stroll, not in a park where he could admire individual flowers, trees or birds but on a school playing field. He likes to walk across the field to the edge, where it drops 20 feet, and then back again – a routine that recalls for him his more epic hikes over the North York Moors.

The emptiness and openness of moorland and the dominance of the sky seem to facilitate communion with the Creator. Gerald McGuire: *'It's almost a religious experience. I talked about the wide open landscape and the sky, and there's a sense of God being there, Who made it all. It's spiritual in a very big way.'*

Asymmetry and homogeneity. Wilderness landscapes must have no obvious pattern, but at the same time they must be simple. Moorland devotees like to see long ridges with smooth lines unhampered by objects such as trees and woods, let alone man-made artifacts like electricity pylons. These qualities and the absence of the variety of birds, flowers, trees and buildings found in lowland landscapes make moorland repellent to many ordinary people. As a retired miner from Rotherham in Yorkshire put it to me after he had taken a holiday trip round Exmoor: *'The moors – they don't appeal. You've seen one bit of heather, you've seen the lot. Too much wide open spaces.'*

Tom Stephenson, Sylvia Sayer, Gerald McGuire and Malcolm MacEwen are uninterested in individual plants and animals when they are out on the moors. Spectacular birds like ravens, merlins, hen harriers, even peregrine falcons could flit past unnoticed or at least undistinguished. Nor do they bend down and admire, close up, a sprig of heather or gorse, lichens, mosses, fungi or a blade of grass. How then do these people perceive the living creatures of the moors if they do not look at them individually?

Birds are part of what Malcolm MacEwen calls *'the silence with sounds'* of the moors. The curlew and the buzzard, through their weird cries, are perceived as background noise, a feature of the moors rather than of the birds themselves. Flowers are perceived *en masse* through their smell and through their colour. Although all five of my enthusiasts are entranced by the richness and changing colours of the moors, in particular the brown and purple of the dying bracken and the heather in autumn, colour is not a primary condition for 'wilderness' since these people love the moors at all seasons of the year.

Emily Brontë, writing in December 1838, explains in a poem why she felt compelled to wander on the bleak Pennine moors around Haworth Parsonage even in midwinter. She writes:

> How still, how happy! Now I feel
> Where silence dwells is sweeter far
> Than laughing mirth's most joyous swell,
> However pure its raptures are.
>
> Come, sit down on this sunny stone;
> 'Tis wintry light o'er flowerless moors –
> But sit – for we are all alone
> And clear expand heaven's breathless shores.

In these lines, she encapsulates something of the spell the moors cast over their admirers: the wide open spaces, the silence, the solitude. Although

Sylvia Sayer thinks that the sight and smell of Dartmoor dressed in her lovely purple heather is breathtaking, she actually prefers the moors in winter because there are fewer holidaymakers around.

Height. For devotees of wild country, height is what distinguishes moorland from lowland heath. At first sight, lowland heaths such as those of Dorset, Surrey and the Suffolk Coast share many of the characteristics of moorland landscapes: monotony of vegetation – mainly heather and gorse – wildness, openness and silence. Yet none of the five people I talked to knew any lowland heaths well, let alone liked them: in general they were repelled by them. The main reason that the heaths have no hold over devotees of moorland is that they are low-lying. 'Wilderness' seekers enjoy the muscular activity and physical exertion involved in their climb up to the moors and in changes in level once they are up there. Further, it does not seem unreasonable to suppose that the moor-lover's climb to the wilderness establishes an aloofness from his fellow men below, which helps foster the sense of in- dividuality he seems to seek.

Freedom to wander at will. Devotees of moorland exult in the liberation they feel on the moors – liberation not only from imprisonment in the towns where most of them spend most of their lives, but also from the confinement of the lowland countryside, where public access is largely restricted to predetermined routes. They like to feel able to roam wherever the mood takes them. They may well be aware that, like the 'naturalness' of the moors, this freedom is illusory, but it is no less important for that. Malcolm MacEwen: *'The fact that you can walk where you like is in fact quite unrealistic in a way, because moorland is generally not very good walking: it's full of bogs and you find that when you move off the track it's quite hard going unless you know how, so you wouldn't. But the sense that you're free and that if you wish to you can move where you like is to my mind enormous.'*

A track is an affront to freedom of movement: it is also usually evidence of human handiwork. Thus lovers of wilderness exult most in a completely trackless piece of moor.

The absence of human handiwork. 'Wilderness' must appear 'natural' and untouched by Man. It may have witnessed many activities, such as battles, over the centuries, but it must seem to have survived unblemished and unaltered since its creation. Devotees of 'wilderness' may be well aware that the naturalness they admire is spurious because the landscape is the product of human activity. But it's the appearance that counts. *'I emotionally and instinctively regard moorland as being natural,'* says Malcolm MacEwen, *'but technically I know that it's clearly not the right term. I feel it in that way, as being untouched landscape although I know this is in fact to a certain degree an illusion.'*

Secondary conditions

Besides the necessary conditions, a number of other characteristics may heighten the appeal of a piece of land to lovers of wilderness, although none of these secondary attributes is essential. Nor is any combination of them sufficient to qualify an area as 'wilderness' in the absence of the seven primary conditions. The secondary characteristics include the possession of relics of ancient man, undulation, wind, and the absence of human beings not fully appreciative of the role of 'wilderness'.

Relics of ancient man. Not all wilderness seekers feel the need to cut themselves off from all traces of their species. Whereas evidence of present-day man – in the form, for instance, of ice-cream vans, litter, nuclear power stations and country buses – seem to be able to destroy 'wilderness' in a landscape which meets all the primary conditions, relics of prehistoric man can actually enhance 'wilderness'.

Dartmoor more than any other British wildscape reeks of history. The archaeologist Jacquetta Hawkes has said that Dartmoor ought to be thought of as one great ancient monument. Dartmoor was the earliest home of man in Devon: it has been occupied for 40 centuries since Early Bronze Age man, in a kinder climate than that of today, grew corn and reared animals on the moor. Traces of past civilisations in the form of hut circles, rings of standing stones, cairns, barrows and mediaeval village remains abound on Dartmoor. It has been estimated that 50 per cent of Dartmoor's ancient monuments have not yet even been mapped. For both Sylvia Sayer and Kate Ashbrook, the discovery of previously unrecorded ancient monuments, the monuments themselves, their preservation, and the strong link with the past that both the monuments and the landscape itself evoke are central features of their love of Dartmoor.

Prehistoric remains are at their most visible on moorland when the bracken has died down, and Kate Ashbrook loves to wander on Dartmoor in winter even though it is bleak, cold and black, largely because of the thrill she experiences in discovering prehistoric remains. For Sylvia Sayer, the ancient monuments are the most interesting feature of the moor: wherever she goes on Dartmoor she finds herself searching for traces of past activity. What feelings do these ancient remains inspire? Sylvia Sayer: '*I just love every one of the ancient monuments and feel they have this great fascination: there were your forerunners and they lived on the moor, they could wring some sort of living out of its soil.*' Kate Ashbrook feels an even closer link with the actual people who lived and worked on the moor in centuries past: '*Often when I'm walking alone I think of ancient man and I stand on the hillside, perhaps near a settlement, and I think he looked out from here and he loved this view. I do feel that they appreciated where they were and they built their homes in certain places because they liked it. I often stand there and I look across to other settlements on other hills and think how they signalled to each other. It's terribly interesting and you do feel a real link and you know that perhaps you're feeling much the same as they did in appreciating it.'*

Wind. As we have seen, the 'silence with sounds' of the moors is an integral part of their spell, and one of these sounds is the wind. For Anne Brontë, at least, the sound of the wind enhanced the wildness of the moors. She wrote in 1836:

> For long ago I loved to lie
> Upon the pathless moor,
> To hear the wild wind rushing by
> With never ceasing roar.
>
> Its sound was music then to me,
> Its wild and lofty voice
> Made my heart beat exultingly
> And my whole soul rejoice.

The absence of unsympathetic people. As a breed, wilderness lovers revel in solitude. *'I've been alone most of my life',* says Tom Stephenson. *'I like to walk alone and feel I'm alone in the world. And I can get that on the moors more than anywhere else.'*

However, solitude is not absolutely essential: the necessary experience seems to be obtainable in the company of a spouse or close friend. What does seem to be important is the absence of unappreciative people or people who appreciate the moors for the 'wrong' reasons. If such people are noisy, untidy, or numerous, so much the worse.

Why preserve wilderness?

'Inspiration', particularly for townspeople, is the main reason Britain's champions of wilderness advance today to justify the protection of moor and mountain.[3] Gerald Haythornthwaite, an indefatigable campaigner for wilderness and for many years chairman of the Standing Committee on National Parks, put it like this:

> Man has need of direct personal relationship with his natural sur-
> roundings in which he can enjoy the grandeur and the richness of land
> and sea, and feel the force of the elements. A man is only half a man who
> cannot exult in a storm on a moor, or a mountain top, or in the sea, or be
> enraptured at the sight of a brown squirrel on the garden wall, or a fox
> in the field. Without such things I believe we shall lose contact with the
> source of all fresh inspiration.[4]

As something 'natural' and as the antithesis of the man-made world, wilderness provides a perspective on city life and the human condition more generally. Gerald Haythornthwaite:

> To find our true unaverage status, the unique importance that each individual possesses but which the world denies . . . we must have places where we can withdraw and be remote from men and their material works and be enfolded by the natural order of things, able to feel that one can go back to the start and unravel the false conclusions of this and other ages.[5]

Wilderness does not even have to be visited for human beings to draw strength from it. For instance, Sylvia Sayer believes that even though people may not go to wild areas something in human nature benefits from knowing that they are there. Comments Kate Ashbrook: *'Even if I'm away from it, just knowing Dartmoor is there is terribly important. It's because it's natural and basic, a raw material that everyone needs from time to time . . . Dartmoor is so uncomplicated. It doesn't need anything material, nothing beastly and man-made to make it the way it is.'*

However, moorland has more practical uses. Walking on the moors is difficult: unexpected peat bogs lie ready to trap the unsuspecting hiker. Even with a track, the rambler has to keep his wits about him to maintain any sense of direction, particularly in mist. The challenge that moorland walking provides, for instance for children striving for their Duke of Edinburgh awards, is a secondary justification wilderness lovers advance for its preservation. Kate Ashbrook: *'The challenge is awfully important because life today, for the young included, has become far too soft and I think it is very important to have challenges left, so long as they don't do it in such numbers that it ceases to be a challenge at all because there's so many of them.'* For Tom Stephenson, the sheer pleasure and physical well-being moorland walking can give are the main reasons for preserving the moors – and no less important in his eyes, for opening them up for public access. He fought for many years to secure access to the Forest of Bowland so that it could become a great playground for Lancashire folk.

Sylvia Sayer does not advocate the preservation of wilderness solely for the benefit of human beings, however. Like Frank Fraser Darling,[6] she believes man has a duty to preserve wildernesses for their own sake. So she believes all nature's wildernesses – even the vast tundra – should be preserved, unlike other conservationists, such as Malcolm MacEwen, who feel that the position of a tract of wild country largely determines it claims to preservation: he feels, for example, that Exmoor is important because moorland is scarce in the southern half of England.

The shaping of public policy

1949–55: clean sweep of the parks. Enthusiasm for moorland on the part of a small group of people of whom Tom Stephenson, Lady Sayer and Gerald McGuire were three, dominated the post-war disposition of the countryside. All ten of our national parks, our most protected large landscapes, contain moorland. For more than half of them – Northumberland (which includes the Cheviots), the Brecon Beacons, the Peak District, Exmoor, Dartmoor, the Yorkshire Dales and the North York Moors – the presence of heather or grass moorland was one of the main reasons why the National Parks Commission selected them. However, such evidence as there is – for instance Lowenthal and Prince's survey of English landscape tastes[7] – suggests that the passion for moorland of those such as Tom Stephenson and Lady Sayer is not shared by the mass of the people, who prefer gentler landscapes. So how did the lovers of moorland gain their ascendancy?

Parliament decreed in the National Parks and Access to the Countryside Act of 1949 that national parks should be selected for their natural beauty and the opportunities they afforded for open-air recreation, having regard both to their character and to their proximity to centres of population. This might seem to suggest such areas as the Cotswolds, the Chilterns, the Weald, the North Downs and the Dorset Downs as candidates. None of these areas has been selected or even discussed seriously during the last 50 years. Our national parks enshrine only three significant landscape types – moorland, mountain and cliff-top. There is no chalk downland, wealden landscape or fen country in any national park, nor any sizeable stretches of coastal marshland. Our lowland vales with their patchwork of field and hedge, down and wood, spinney and stream – highly prized by many people, especially overseas visitors – are nowhere specifically protected by national park designation. What is more, despite the Act's emphasis on accessibility to large centres of population, there is still no national park near London, Southampton, Birmingham or Bristol. Instead, most of our parks are far from rather than close to our largest towns and cities.

A 1971 survey by the *Geographical Magazine*[8] has provided the only indication so far of the general public's preferences for new national parks. The magazine asked its readers: 'If you had a choice, which other areas would you designate as national parks?' The areas put forward by the 3000 readers who replied were ranked in order of preference. The Cotswolds came out top; eight of the top ten proposals were in lowland Britain and included such areas as the South Downs, New Forest, Norfolk Broads, Dorset Coast, Chilterns, Weald, and the North Downs of Kent and Surrey.

One reason for the upland bias of our national parks seems to be that the countryside establishment – the people in a position to influence the choice of national parks – was dominated during the 1930s, '40s and '50s by lovers of wild country. At the time, there was no counter-lobby promoting the claims of the lowlands.

It was not just that most people active in the countryside movement during these years preferred mountain and moorland: some of them also considered that these landscapes would 'do people more good' than other types. Professor G. M. Trevelyan, for example, wrote in 1931:

> Nature, no doubt, acts as a comforter and giver of strength even in southern woodlands and on smooth hillsides. But to many of us the moorland and the mountain seem to have more rugged strength and faithfulness with which in solitude we can converse and draw thence strength and comfort. And the mountain above all seems to have personality which says to us as we gaze on it at evening from the valleyhead below – I know, I understand. Such is the lot of man. I watched him through the ages. But there is a secret behind. It will always be a secret.[9]

During the 1930s, Vaughan Cornish was the most prolific writer on landscape aesthetics and a leading conservationist. A typical child of the Romantic Movement, Cornish thought that the supreme forms of landscape were 'the mountain peak soaring to the clouds' and dramatic sea-cliffs, 'The bold headland wreathed above in driving mist and drenched below by the spray of battering waves.'[10] He was not interested in lowland agricultural scenery, which was redeemed in his eyes only by the element of drama provided by buildings, especially church steeples.

Undoubtedly the most influential figure in determining the type of countryside that came to make up our national parks was John Dower. It was his Government-commissioned report on national parks,[11] published in 1945, which was to prove the key to the shape of Britain's national parks. John Dower was born and brought up at Ilkley and lived in Northumberland and the Yorkshire Dales. He was a hill-walker, preferring wild moor and mountain to lowland countryside. The ten areas he put forward as candidates for priority designation were all dominated by mountain, moor and sea-cliff, particularly moor; understandably so, since he stated unequivocally in his definition of a national park that national parks should be confined to 'wild country'. Dower justified this restriction on the grounds that it is only in wild country that the public either wants widespread access or can be given it, but it seems that his decision was also influenced by his personal preference for wildscape.

Shortly after the Dower Report was published, the Government appointed a committee chaired by Sir Arthur Hobhouse, to consider and report on Dower's recommendations.[12] Anxious that national parks should cover a wider variety of landscape than Dower had suggested, the Hobhouse Committee proposed two more areas – the South Downs and the Norfolk Broads.

This move came to nothing. The Broads were ruled out by the National Parks Commission as unsuitable. When the commission first discussed the South Downs, Tom Stephenson, who was then a member, moved that they be rejected because they were not wild enough; the commission then shelved the idea of designating them. When, in 1956, the possibility of designating the Downs was revived, so much of the rough grass had been ploughed up that the commission thought them no longer suitable: no large areas were left over which the public could wander at will, in their view an essential requirement for a national park.

Having dropped the Broads and the Downs, the National Parks Commission did not replace them with other areas representing types of landscape other than wild country. Instead, they added to Dower's list the high, wild moors of the Cheviots. The inclusion of these hills in the Northumberland National Park owes more to Tom Stephenson than to anybody else. Stephenson, who was on the Commission between 1949 and 1953 when six of our ten national parks were designated and when negotiations started for a further three, is convinced that wild country is the *sine qua non* of our national parks. Like all but one of the conservationists I interviewed, Tom Stephenson referred to John Dower and to 'wild country' when asked to define a national park – although Parliament omitted this phrase from the definition in the 1949 Act.

Four other factors were also at play that made the decision-makers turn their eyes almost unhesitatingly to the uplands when establishing Britain's national parks. There was the precedent of North America, where national parks were established in the 1870s and have always been used to preserve, and open up for public access, spectacular and wild landscapes. At the same time as the first national parks were being designated in the United States, a campaign to secure public access to mountains began in Britain. This campaign was later widened to embrace moorland (during the 1930s the campaign was focused on the grouse moors of the Peak District which were then barred to walkers) but it did not take in other types of landscape feature such as woodland. In addition, there was the prevailing idea that national parks should be areas where the public could roam freely, an idea which started with Vaughan Cornish and was supported by Dower. Lowland farmed countryside was therefore ruled out not because it lacked appeal, but because access was mainly confined to footpaths.

There was, however, another powerful reason for omitting lowland vales and chalk downland from the list of national parks. During the early postwar years in particular, the importance of food production loomed large in the legislators' minds. The wartime threat posed to our food imports by German submarines had demonstrated the strategic importance of our farmland. In 1949, the acid and poorly drained moors of the north and west looked bad prospects for agriculture, but the chalk downland of the south-east had been shown to have enormous potential: large tracts had been ploughed up in wartime 'reclamation' schemes. These four factors combined to make it seem beyond question that lovers of wilderness should see the areas they most loved consecrated as national parks.

1955–80: unquestioned priority. From the mid-1950s until 1980, the moors received priority in almost all discussion of countryside protection. National parks have always been considered our most important countryside, and since the term 'national park' has become synonymous with moorland, this priority was reinforced. For example, the Ministry of Agriculture, Fisheries and Food decided in 1980 to alter the arrangements governing the issuing of capital grants to farmers for schemes to increase output such as wetland drainage, the erection of fences, the conversion of rough grassland to ryegrass pasture and so on; but for the national parks, and therefore moorland, special procedures were introduced to increase the prospects for conservation of areas of land for which capital grant was being sought.[13]

The thinking of the 1960s, enshrined in the Countryside Act 1968, moulded countryside recreation provision in England and Wales for the following 20 years, and one of the main objectives of the policies devised in this period was the need to protect moorland for the wilderness seeker.[14] The mood of the period was set by architect and planner John Dower in a treatise entitled *The Fourth Wave* which prophesied a large increase in the numbers of car-borne visitors to the countryside and proposed that instead of allowing them to spread throughout the country, they should be concentrated in sites where they would do little harm. 'We must discriminate,' said Dower, 'fitting each feature and region to the recreation it can best satisfy, gathering the crowds into places which can take them, keeping the high, wild places for the man who seeks solitude.'[15]

The mid- to late-1970s did, however, see a questioning of the automatic priority to moorland. In 1973, the Government commissioned the first study of the national parks system that had been established 24 years earlier, and the Sandford Committee, reporting in 1974,[16] noted the uneven distribution of the parks and their overall lack of landscape variety. To remedy these imbalances, the Committee recommended that the Countryside Commission should examine more diverse types of landscape when considering possible new parks, and should seek to redress to some extent the uneven geographical spread of the existing parks. The Government supported this recommendation in 1976.[17]

The Countryside Commission responded in 1977 by asking for public reaction to the suitability of national park status for the Norfolk Broads. This proposal was later shelved because of opposition from the local authorities concerned. The Commission is not at present scouring the lowlands for other candidates for national park status, and shows no likelihood of doing so in the future.

Nevertheless, the Sandford Committee did also put forward a recommendation which, had it been implemented, would have radically strengthened the protection afforded to moorland (and mountain) at the expense of other landscape types. The Sandford Committee proposed that the wildest heartlands of our existing national parks should be specially designated as 'national heritage areas'. This new designation would replace national park status as the highest form of protection available to landscape in England and Wales. Within the new areas, the majority of which would almost certainly have consisted of moor and mountain, no development would be permitted without Parliamentary approval and the entry of vehicles would be strictly limited. The idea behind the national heritage area was not new: the United States Wilderness Preservation Act, 1964, provided for the establishment of 'wilderness areas' in addition to the system of national parks, and 127 such areas had already been set aside by the end of 1976. Nearer home, the national heritage area concept resembles closely a proposal made in 1973 by the journalist Jon Tinker that 'wilderness areas' be established in the wildest parts of moor and mountain in our existing national parks and that these areas should have more protection than other landscapes.[18]

The national heritage area idea reappeared in 1979 with a proposal by the Countryside Review Committee, a group of civil servants with most responsibility for the countryside, that the system of national parks and areas of outstanding natural beauty should be replaced by a system of first-tier areas and second-tier areas.[19] This proposal has also been formally turned down by the government, but it does reflect an approach which still finds favour in many official circles. The first-tier areas that the Review Committee envisaged would have occupied only between two and three per cent of the land area of England and Wales, and development would not have been permitted without Parliament's approval. The Review Committee, however, did believe that first-tier areas should be selected from within the present AONB areas as well as the national parks and in a few cases from outside both.

1980 onward: hegemony challenged. 1980 seems to have been a watershed as far as the hegemony of moorland is concerned. Although the Conservatives' 1980 Wildlife and Countryside Bill, like its aborted Labour predecessor, confined its landscape conservation provisions to measures to increase the protection afforded to moorland in national parks, the debates on the Bill and the amendments tabled reflected a sudden and growing concern with Britain's long-neglected lowland countryside. One amendment, for instance, tabled by Lord Winstanley, sought to extend the principle of advance notification for the ploughing of tracts of moorland in national parks covered by orders under Section 14 of the Countryside Act[20] to a range of lowland as well as upland landscape types. Under the terms of the amendment, which was not adopted, any owner or occupier who wished to convert to intensive agricultural land a tract of mountain, moor, heath, down, cliff roughland, foreshore or woodland anywhere in the country would have had to notify his local planning authority before he went ahead.

An amendment tabled by Lord Melchett, the opposition's chief spokesman on the Bill, and by Baroness David, sought to extend the scope of development control to embrace the removal of the features that characterise the ordinary, basic countryside of England and Wales: hedgerows and hedgerow trees, woodlands and marshes, streams and ponds, downland and heathland as well as moor. Lord Melchett's amendment followed a period of considerable public debate about conservation in which a number of important trends emerged. One of these was the growing belief that the countryside as a whole serves a wide range of important functions, and that it is this countryside, not just wild moorland in national parks, that matters most to most people. Professor Sir Colin Buchanan put this point well in a letter published in *The Times* on 18 November 1980. He wrote: 'Surely the point about the country-side is that it is used for so many different purposes. Farming is one; looking at it from cars or trains is another. It is also used for exercise, for rambling, for camping, for riding, for adventure-training, and it provides source material for artists, poets, biologists, ornithologists, zoologists, archaeologists, architects, historians and many other people.'

It now seems likely that the moors will face competition for their privileged place in the hierarchy of landscape features meriting protection. In future, they are likely to bid for support alongside chalk downland and lowland vale, river valley and woodland, coastal marshland and fen country. In this competition, they may lose their long-held supremacy, but they should be able to hold their own. For there is now real popular enthusiasm for the moors even if it is not sufficient to justify their past privileges. On Ilfracombe promenade in August 1978, I asked a few holidaymakers how they felt about Exmoor. A nursing sister told me it was the moors' aura of history that appealed to her: *'They've been there untouched for hundreds of years; people have been fighting battles over the same moors.'* An electricity board showroom manager with whom I spoke loved the natural life of the moors; a Chinese student relished the sense of freedom – *'it broadens your heart . . .'*; a retired miner from Rotherham liked the individual heather plants; a young factory girl spoke of the peacefulness of the moors. For the visitors to whom I talked, the colours of the moors were the most generally popular feature, and it is this feature that makes the attractions of the moors for Britain's artists understandable. There are few more breathtaking sights anywhere in the world than Dartmoor in August, aflame with a mosaic of golden gorse and purple heather, ripening moor grass and the russet hues of dying bracken. Quite apart from their other intrinsic charms, the moors are also a vital ingredient in the whole landscape mix of England and Wales. Their rough open spaces provide a contrast with the gentle patchwork of fields and hedges below that emphasises the pastoral intimacy of our typical lowland countryside.

The moors do have conservation needs. Rich deposits of mineral ores underlie many of our moors and quarrying can damage the landscape. Road building and road widening, house and factory building threaten parts of the moors; and the Army's occupation of a third of Dartmoor has caused much damage, particularly to ancient monuments on the moor, and has necessitated restrictions on public access. As in Britain as a whole, however, the impact of activities like these is dwarfed by that of agricultural change, though this has affected moorland less dramatically than other landscape types up till now. It has been estimated that Britain's moors have been enclosed and reclaimed for intensive agriculture at an average rate of 12 000 acres a year since the war.[21] Because the original acreage of moorland was so great, the impact of this annual rate of change has been less obvious than the impact of agricultural change on other landscape types.

Nonetheless, it has already had far-reaching implications for landscape, access and wildlife in particular places and threatens to grow into a far larger problem in future. The enclosure of moorland and its conversion to ryegrass monoculture for feeding animals has already eaten away one fifth of the moorland of Exmoor since the war; and nearly 60 square miles of moor, most of it with *de facto* public access, were fenced off and 'improved' for intensive agriculture or planted with conifers in the North York Moors National Park between 1950 and 1975.[22] Both these activities do, of course, destroy completely the fragile sense of wilderness of the moors. Enclosure for agriculture destroys the freedom to wander at will as well as the point of it. The sense of timelessness evaporates when the 'silence with sounds' of curlew and buzzard gives way to the drone of farm and forestry machinery. And the conversion of the wilderness to a food or timber factory may seem to some wilderness lovers more offensive than the fouling of Man's own nest.

It is forestry, however, that looks set to pose the greatest threat to our moors over the next 50 years. Although the government has not yet announced precisely what scale of new planting it will encourage, two reports point the way. All that would be needed from the government for a major expansion in forestry would be a commitment to guarantee the present level of tax advantages and planting grants for private foresters coupled with some support for the Forestry Commission. For forestry, like agriculture, may take more or less all the land it wants, since the industry is exempted from the need to seek the consent of the community as a whole through the planning process before the environment is changed. The Forestry Commission's 1977 report, *The Wood Production Outlook in Britain,* suggests that it is reasonable to assume that 1700 square miles of upland country in England and Wales alone will be planted within the next 50 years: this would mean that one third of the existing open moorland in England and Wales will be covered in plantations within the next century. The Centre for Agricultural Strategy's report, *Strategy for the UK Forestry Industry,* published in 1980, forecasts a 50 per cent increase in UK consumption of wood and wood products between 1980 and the year 2000, and a 90 per cent increase between 1980 and 2025. As it predicts that this massive increase in demand will take place at a time when the world price of timber is likely to increase, the report suggests that the maximum feasible rate of planting in Britain to meet this demand is an extra 185 square miles every year between 1980 and 1990, leading to an extra 7500 square miles more land covered in forestry plantations (largely conifer) by 2030. If planting on this sort of scale goes ahead, and agricultural intensification continues to bite into other areas of moorland, then the threat forestry poses to the moors may come to match that which agriculture poses to the lowlands. Uplands and lowlands may then have to face together a similar challenge. If we allow the challenge to prevail, the landscapes that will greet our grandchildren in the next century will be vast prairies of cereals or grass monoculture in the lowlands and, in the uplands, great timber factories. It is right that the moorlands should share the privileges which have caused other landscapes to be deprived of the help they have needed more urgently than the moors. Yet it would be sad indeed if this process were to be accompanied by a real loss of interest in moorland just when it is coming to need the attention it has enjoyed for so long.

Notes

1 Porchester Report 1976. *A Study of Exmoor*. London: HMSO.
2 For an account of the history of American attitudes to wilderness, see Nash, R. (ed.) 1972. *Wilderness and the American Mind*. New Haven, Conn.: Yale University Press.
3 Lady Sayer develops this theme in *Wild Country: National Asset or Barren Waste?* 1970. Standing Committee on National Parks of the Councils for the Protection of Rural England and Rural Wales.
4 Haythornthwaite, G. 1966. *My Case for Preservation*. London: Council for the Protection of Rural England.
5 Haythornthwaite, G. 1960. *The National Park Dilemma*. London: Youth Hostels Association.
6 Darling, F. F. 1970. *Wilderness and Plenty*. The Reith Lectures. British Broadcasting Corporation.
7 Lowenthal, D. and H. C. Prince 1964. The English landscape. *Geog. Rev.* **54**, 309–46 and idem 1965. English landscape tastes. *Geog. Rev.* **55**, 186–222.
8 Board, C. and R. Morgan 1971. Parks for people. *Geog. Mag.* **43**, 640–4.
9 Trevelyan, G. M. 1931. *The Call and Claims of Natural Beauty*. The Rickman Godlee Lecture, 26 October.
10 Cornish, V. 1930. *National Parks and the Heritage of Scenery*. London: Sifton Praed.
11 Dower, J. 1945. *National Parks in England and Wales*. Cmd 6628. London: HMSO.
12 *Report of the National Parks Committee (England and Wales)* 1947. Cmd 7121. London: HMSO.
13 Ministry of Agriculture, Fisheries and Food 1980. *Agriculture and Horticulture Development Regulations*. London: HMSO.
14 For a detailed report of the way in which this thinking was translated into policies see my article (1979): Metropolitan escape routes. *The London Journal* **5**, 87–112.

15 Dower, M. 1965. Fourth wave. *Architects' Journal*, 20 January.
16 *Report of the National Parks Policies Review Committee* 1974. London: HMSO.
17 Department of the Environment and the Welsh Office 1976. *Report of the National Parks Policies Review Committee*. Circular 4/76. London: HMSO.
18 Tinker, J. 1973. Do we need wilderness areas? *New Scientist* **60,** 42–4.
19 Countryside Review Committee 1979. *Conservation and the Countryside Heritage*. London: HMSO.
20 Section 14 of the Countryside Act 1968 empowers the Secretary of State for the Environment to make orders covering areas of national park. Any person intending to plough up unimproved moorland in an area covered by an order must first notify his national park authority of his intention to do so. The authority may then seek to preserve the moor in question through land acquisition or through a legal and financial agreement with the landowner, which almost always involves their paying the landowner compensation.
21 Parry, M., A. Bruce and C. Harkness 1981. The plight of British moorlands. *New Scientist* **90,** 550–1.
22 For a detailed account of the impact of agricultural change on moorland in England, see my book (1980): *The Theft of The Countryside*. London: Temple Smith.

5   *Revisiting valued landscapes*

DAVID LOWENTHAL

Any valued locale is *ipso facto* memorable; we revisit it, recalling or copying its real or imagined lineaments so as to preserve and heighten previous experience. Scenes from our own past and from historically remote times become 'a living link between what we were and what we have become,' in Margaret Drabble's words.[1] The tides of taste for moors and mountains, meadows and pastures, castles and cathedrals and everyday locales continually shift.[2] But whatever they are and however they are experienced, favoured scenes linger on in memory. Here I examine the significance of such recollections and imagination.

I have elsewhere set forth the bases of attachments to the past – notably the desire for evidences of antiquity, duration, accretion, continuity, and identity.[3] In this essay, I first identify some specific virtues attributed to bygone scenes, whether revisited in reality, re-experienced in surrogate form, or simply recalled in imagination. I go on to discuss favoured routes to these landscapes, chosen modes of access to remembered and historic places. Finally, I touch on the difficulties and drawbacks of trying to revisit and recreate the past.

The virtues of bygone scenes

Much of the pleasure taken in relic-strewn or reminiscent landscapes derives from their apparent contrast with the scenes that now surround us. The belief that things were better back then is widespread, especially among the elderly, whose preferred landscapes often antedate their own arrival on the scene. W. G. Hoskins adjudged the English countryside of 1500 'infinitely more pleasant a place,' with 'plenty of fresh air and space for everybody,' than the mid-20th century's Nissen huts, arterial bypasses and murderous lorries. Especially since 1914, 'every single change in the English landscape has either uglified it or destroyed its meaning, or both'.[4] Hoskins would preserve such relics as the county of Rutland, still 'a human, peaceful, slow-moving, pre-industrial England, with seemly villages, handsome churches, great arable fields, and barns,' and at each county entrance affix a notice: '*Human Conservancy: Abandon the Rat-Race at This Point.*'[5]

Many associate such virtues with the time of their own childhood. As an old man, E. M. Forster recalled the English landscape of his youth:

There was a freshness and out-of-door wildness in those days which the present generation cannot imagine. I am glad I have known our countryside before its roads were too dangerous to walk on and its rivers too dirty to bathe in, before its butterflies and wild flowers were decimated by arsenical spray, before Shakespeare's Avon frothed with detergents and the fish floated belly-up in the Cam.[6]

Forster's paradisaical image merges the remembered with the legendary past; his 1907 landscape presumably differed little from Shakespeare's England, whereas 20th-century change seems massive, disjunctive and vile. But the supposed superiority of the quite recent past is no new belief. As Raymond Williams has shown, successive 'olde Englands' have for centuries been beloved of every later age: retrospect reshapes and gilds a supposedly happier time, linked with countryside childhood.[7]

Attachment to an idealised past merges with English chauvinism. Back in a Suffolk pub after an American sojourn, two Englishmen meet 'an affable old countryman' who

had never in his life been more than five miles away from the village. Nevertheless, 'Back home again, eh?' he cried. 'Ah, you got to go a long way to beat old England, eh?' and his words set up in our minds a train of thought that culminated in the concept of 'phogy' . . . that stout spirit of British phlegm . . . in which the artificial stands superior to the real, the traditional to the new, . . . and makes living in England like walking through syrup. The phogy is half fogy, half phony – phony because there is a structure of tradition and pretense in the phogy. . . . Add a touch of the word 'foggy' and you have our meaning. It was natural to be nineteenth century in the nineteenth century, and anyone could do it, but in the twentieth it takes quite a lot of toil.[8]

Outdated scenes and artifacts are no less treasured for demanding effort. 'Wouldn't a spin dryer be more effective' (than a 19th-century mangle)?, asks a guest in Penelope Lively's *Treasures of Time*. 'I've got fond of it,' her hostess replies; 'if a thing is nice to look at *and* reasonably functional – *and* old – then isn't it worth sacrificing a bit of convenience?[9] Indeed, the inconvenience of age – waxing an old car, restoring an old house, finding appropriate period accessories – enhances the pleasure taken in old-fashioned artifacts and customs. The ritual of preparing, serving and consuming afternoon tea, for example, harks back to an age of grandparents, of servants, of leisure to sit and sip. The sense of gentility lends charm to the ritual whether or not an actual past is recalled by it. 'It's posh to like old things', says a knowing youngster in Lively's *Road to Lichfield*. 'Antique furniture and houses with beams everywhere, and vintage cars. And old maps. Dead posh. It shows you've got nice taste.'[10]

The sounds and smells of yesteryear engender similar nostalgia. 'In the past the trains either whistled more or we heard them better', reminisced one old lady about the 1920s: 'They had more personality'.[11] Correspondents to *The Times* bemoan the demise of old sounds. 'Alas! we no longer hear the clip-clop of the horse with the milk-cart in the early morning'; 'it was tragic that the modern child could never hear the joyous scream of the "puffer-train".'[12] Another writer recalls laundry cupboards stuffed with fragrant sprigs and little sachets, vapours from hot flat irons at the tailor's, mulled wine filling the air with steaming cinnamon, the smell of cut logs at the sawmill, bluebell woods and bonfires at dusk, 'scented stigmata of an age that has been cast for ever into oblivion; . . . all that is left is a desolate, hygienic emptiness.'[13]

The valued past is a heterogeneous medley of scenes, objects and thoughts. A putative English heritage exhibit for schools features Roman lamps, coins and tiles; facsimile pages from Bede's *Ecclesiastical History,* the Caxton Bible, the Lindisfarne Gospel, the earliest edition of Chaucer, the Shakespeare First Folio, and the Paston letters; Celtic metalware; a Viking shield, buckle and sword; and photographs of vernacular architecture and domestic utensils. But such nationally significant relics form only the tip of the popular iceberg. Iron Age forts, Greek temples, Celtic jewellery, Georgian buildings, neo-Gothic whimseys, traces of Roman roads, ridge-and-furrow plough marks, steam engines, windmills, Morris dancing, Macaulay's histories, table mats and glassware with 18th century scenes, old paintings fake or real, Art Deco cinemas, coronation mugs, 1930s juke boxes, 1950s dress styles, 1960s movies, runic inscriptions, Golden Oldies, the treasures of Tutankhamen and the terrors of Madame Tussaud, childhood memories and chats with grandma, seaside souvenirs, family photographs and family trees, old trees and old money, preserved locks of hair – all these connect us with a valued past.

The English are not alone in preferring olde England; foreign visitors deeply venerate its antique flavour. 'The *prime* reason we come', one American explains, 'is to experience what is left of the past'.[14] Trekking from the Tower of London and Westminster Abbey to Stratford and Oxbridge, York, Canterbury and Bath, they seek out an England devoid of the 20th century or even the 19th. Here, as one of Henry James's visiting Americans describes collegiate Oxford, are 'places to lie down on the grass in forever, in the happy faith that life is all a vast old English garden, and time an endless English afternoon.'[15]

So deep-seated is the desire to relive the past that emulated historical scenes are apt to be mistaken for the originals. A 1974 urban 'farm' in Kentish Town soon acquired the mystique of tradition; 'people were eager to believe that it was an actual fragment of farmland overlooked for a hundred years and miraculously rediscovered like the Sleeping Beauty's domain'. The farm created 'a sense of the revival of the lost past', lending strength to local faith, in Gillian Tindall's phrase, that 'the fields are not only sleeping underneath: they are *here,* exposed once again'.[16]

Scenes from the past are appreciated for their supposed stability. We contrast the hectic pace of our own lives with the slower tempo of days gone by. Even if most people in the past lived at the mercy of untoward accident and illness, the world in general seemed to alter more slowly. 'Before 1900, things didn't change so fast as now'[17] is a typical folk-version of Henry Adams's famous exponential law about the increasing velocity of history.[18] Victims of 'future shock', we envy our predecessors their more leisurely days and weeks, months and years.

The past also helps us to distance ourselves from the present, to get away from the here and now, to turn back to a time for which we have no responsibilities and when no one could answer back. Even if the past was no golden age, historical scenes can, as Hoskins suggested, alleviate the contemporary rat-race. 'Come to Williamsburg. Spend some time in Gaol', urges a travel advertisement, with tourists shown grinning in the 18th-century stocks: 'it will set you free' – free from day-to-day cares in the humdrum, workaday present. Even a contrived past may alleviate rapid or dislocating change. Rest cures in historically frozen Amish villages, or in the simulated past of Williamsburg or Mystic, could compensate for the pace of modern life. In such islands of time, 'individuals who need or want a more relaxed, less stimulating existence should be able to find it,' Toffler suggests.[19] 'Men and women who want a slower life, might actually make a career out of "being" Shakespeare or Ben Franklin or Napoleon – not merely acting out their parts on stage, but living, eating, sleeping, as they did.' (Indeed, Plimoth Plantation staff actually play the part of particular 17th-century figures – for the benefit of tourists.) Most visitors to Old Sturbridge Village, a recent study shows, found an emotional magic in this 'capsule in time', an 'age-old' enchanting serenity.[20]

Mystery, as well as stability, invests antique landscapes with charm and meaning. An ancient Mexican city was, for Anaïs Nin, a realm of refreshing uncertainty 'rendered into poetry by its recession into the past, . . . allowing each spectator to fill in the spaces for himself'. Fragmented relics lend landscapes an aura of enchantment: 'A city in ruins, as this ancient city was, was more powerful and evocative because it had to be constructed anew by each person'; it was 'never destroyed or obscured by the realism of the present, never rendered familiar and forced to expose its flaws'.[21] The past may become attractive precisely because it is unfamiliar, indeed unknowable.

The past appeals as a land of mystery – strange peoples, exotic customs, magic and witchcraft, splendour and terror – especially to those who know it but little. Historical junkets guided by Costain and Plaidy confirm popular stereotypes that life in olden times was one of intense extremes, glittering and exuberant luxury alternating with cruel and unspeakable misery.

The cherished past often varies from epoch to epoch. The Romans prized ancient Arcadia, the Renaissance admired classical Greece and Rome, Victorian England doted on Greece and the Gothic Middle Ages, and early 20th century Americans fancied their English heritage. Today we enjoy most past periods indiscriminately: the popularity of the whole sweep of history from cavemen to Cape Canaveral is a hallmark of present-day addiction to the past. But as an antidote to today, earlier antiquity often has more appeal than a later one; the remote and malleable past is apt to be preferred to a recent one perhaps painfully recalled. Remoteness from our own time makes antiquity uncanny and mysterious. Walking through the litter of Bronze Age antiquity on Dartmoor gave Sylvia Sayer 'a kind of involuntary recognition that [she had] been here before – a long time before'.[22] An archaeologist uncovering a Roman plumb bob in a dig is thrilled to 'imagine the chap who had last handled it, fifteen hundred years ago, saying "where's that thing gone to now?"'[23] The mystique attached to standing stones and the widespread legend of a ghostly hunt are seen as 'the legacy of unbelievably ancient involvements with place that are so strong that there still lingers an echo of what they were'.[24]

Sheer age lends romance to times gone by, and 'the further those times are removed from us', in Chateaubriand's words, 'the more magical they appear'.[25] A well-known Ridgeway walker views Iron Age features with wonderment that such mighty works could be 5000 years old;[26] a local English museum calls a restorer's landscape model 'a great achievement – and all the more because he did it so long ago'.[27] The 'secrets older than the flood' celebrated in Wordsworth's The Prelude and Shelley's 'thrilling secrets of the birth of time' convey the fascination of the distant and hidden. Uncorrupted by later accretions and experiences, the primordial is pure and perfect. Similarly, we are fond of invoking an ancient rather than a recent authority because the ancestral is 'remote enough to be more manageable in the quest for your own identity – more open to what the heart wants to select or the imagination to remold'.[28] Hence the remote past is a splendid playground for vicarious adventure. 'What makes prehistory interesting is what makes science fiction interesting,' says film director Jean-Jacques Arnaud: 'the ability to dream and create a world.'[29] Very old buildings likewise have a cachet the recent past cannot equal. It has proved hard to preserve the belle époque architecture of Montreux because city fathers 'think that something from 1900 is no more valuable than something brand-new'.[30]

A sense of durability, however, does not necessarily require great antiquity; things that antedate only ourselves may seem to have existed forever. The great ash tree in my garden seems to me immemorial because it was here long before I came, and is clearly older than me. The house next door also antedates my arrival here, though not my birth; consciousness of its relative youth diminishes its aura of permanence. But the wilderness that covers the site of a house torn down two years ago seems 'new' and transient because I vividly recall what was there before.

Continuity is another valued quality of ancient landscapes – the sense of unbroken succession often visible in storied locales. At such places in England 'one treads on three thousand years of history, sometimes four thousand or more'. Hoskins finds it

> satisfying to sit upon a Saxon boundary bank, . . . to know which of these farms is recorded in Domesday Book, and which came later in date in the great colonization movement of the thirteenth century; to see on the opposite slopes, with its Georgian stucco shining in the afternoon sun, the house of some impoverished squire whose ancestors settled on that hillside in the time of King John and took their name from it; to know that behind one there lies an ancient estate of a long-vanished abbey where St Boniface had his earliest schooling, and that in front stretches the demense farm of Anglo-Saxon and Norman kings; to be aware . . . that one is part of an immense unbroken stream that has flowed over this scene for more than a thousand years.[31]

The unbroken stream is a peculiarly English virtue. A community of descent connects the earliest with the latest folk, primeval artifacts with those of today; traces of all epochs survive. Accustomed to domestic continuities, English travellers likewise admire temporal palimpsests abroad. Hence Rose Macaulay took keen romantic pleasure in the historical landscapes of Crete, 'in seeing the Achaean culture imposed on the last Minoan, the Dorian on the Achaean, the Roman on the Hellenic, the Byzantine on the Roman, then the Saracen, the Venetian, the Turkish, and the Cretan of today'. Similarly 'the ghosts of dead ages sleeping together' in the Peloponnese, the 'superimposition of medieval on ancient, modern or medieval . . . thrust up on craggy heights from Byzantine or classical foundations', enrich the mosaic impression.[32]

The accretion of human traces not only adds to knowledge and to romantic sentiment, but enhances the land itself, as in Henry Newbolt's fictional estate:

> Has the slow stream of human life had no effect upon these meadows that it has so long watered? Are they no richer for all this love, no more fertile to the spirit than the raw clearing of yesterday in new-discovered countries? Are there no voices but ours in these old mossy woods and sunlit gardens, no steps but ours by this lake . . .? What, then, is Time, that he should have power to make away with the dearest memories of seven and twenty generations?[33]

Revisiting past landscapes through science fiction

These and other preoccupations with the past are highlighted in the inventions of science fiction, where time travel allows characters to recover scenes of their own youth or of more remote epochs. The immense popularity of this nostalgic genre, ranging from the romances of H. G. Wells to the adventures of *Doctor Who,* reflects a widespread and profound human concern. Indeed, of 27 paramedical trainees surveyed in a recent American study, over half said they would pay substantial sums to recover a year of their personal past, almost one-third to recover a year from remoter history – the men mainly to rework the retrieved period so as to change its outcome, the women to relive past experiences and feelings.[34] Together with public fondness for outdoor history museums and fascination with experiments in Iron Age lifestyles, the speculations of science fiction illumine the nature of and the reasons for modern attachment to landscapes of the past.[35]

Searching for the Golden Age. Mirroring readers' tastes, science fiction writers increasingly express a preference for former times, a belief that the world used to be a better place. Some admire all epochs; 'through the vistas of the years every age but our own seems glamorous', one novelist puts it.[36] On expeditions into the past, a time traveller was forever reminding himself that each world he visited was 'where he *should* be'; any previous date was 'vastly preferable to his own regimented day and age'.[37]

Others fancy some particular period, a specific Eden. The Stone Age delights those who long for a lush, green, unpolluted world, teeming with plants and animals and lightly tenanted, if at all, by mankind. '"How nice it would be"', says a character in Clifford Simak's *Catface,* to travel '"back to the time before the white men came, when there were only Indians. Or back to a time before there were any men at all"'.[38] He elects to live in the Cretaceous, 100 million years ago. The prehistoric world of 15 000 BC seems almost like paradise to the expedition leader of Philip José Farmer's *Time's Last Gift.* Pure air, plenty of grass and water, 'damned few humans, and an abundance of wild life; this is the way a world should be'.[39] Preferring wide-open country with as few people as possible clogging up the scenery, Gordon Eklund's time travellers consider the 20th century less interesting than the 19th, the 19th less desirable than the 18th.[40]

The remote past in Simak's *Catface* is a potential refuge for today's poor and deprived, at the same time relieving modern social pressures. Virgin prehistory would offer pioneers a whole new world of their own. 'Give us the Miocene; we want another chance', cry ghetto dwellers enticed by this vision.[41] Unemployment and crime are curbed in Robert Silverberg's *Time Hoppers* by sending the unwanted back to 500 000 BC – to be eaten by tigers. Others escape bureaucratic oppression by fleeing back to a time before the state became all-powerful.[42]

Some nostalgists opt for a more recent past, the familiar yet still unspoiled era of their grandparents or of their own entrance onto the human scene. The utopian colony in Mary McCarthy's *Oasis* selects a locale 'arrested at the magical moment of [their] average birth-date . . . forty years before, and . . .

at the stage of mechanization to which the colonists wished to return'. The buildings and furnishings of 1910 'took them back to the age of their innocence, to the dawn of memory, and the archaic figures of Father and Mother'.[43]

The late 19th century was by no means perfect, realises the hero of Finney's *Time and Again,* but 'the air was still clean. The rivers flowed fresh, as they had since time began. And the first of the terrible corrupting wars still lay decades ahead'.[44] Gerrold's *Man Who Folded Himself* opts for 'the quiet fifties . . . as early as I dare go without sacrificing the cultural comforts I desire. They are truly a magic moment . . . close enough to the nation's adventurous past to still bear the same strident idealism, yet they also bear the shape of the developing future'.[45]

Enjoying the experience. Recent or remote, the desired past exhibits strikingly similar traits: the world back then is seen as natural, simple, comfortable, yet also exciting and colourful, promising a vital and vivid existence. 'A man could really *live* in this lush, green world', exclaims a time traveller appreciating Illinois in 1856.[46] Though sometimes nostalgic for the modern world, the hero of Robin Carson's *Pawn of Time* realises that 'the old, gray, modern existence' had little to offer compared with 'the new, colorful opulence' of Renaissance Venice.[47] A visitor who 'becomes' Cyrus in ancient Persia finds early warfare more enjoyable than modern foxholes.[48] The 14th-century world was 'cruel, hard, and very often bloody, and so were the people in it', learns a du Maurier protagonist, 'but, my God, it held a fascination for me which is lacking in my world of today'.[49] Early Romans seldom lived long, admits another voyager, 'but while they lived, they *lived*'.[50] A Poul Anderson figure finds his Stone Age mate more compelling than his humdrum 20th century wife;[51] so does du Maurier's onlooker *vis-à-vis* a 14th-century Isolde. Modern New Yorkers seem pallid to Finney's hero in *Time and Again.* 'Today's faces are . . . much more alike and much less alive. . . . There was also an excitement in the streets of New York in 1882 that is gone.' Back then people were 'interested in their *surroundings* [and] carried with them a sense of purpose. . . . They weren't *bored,* for God's sake! . . . Those men moved through their lives in unquestioned certainty that there was a reason for being . . . Faces don't have that look now; when alone they're blank, and closed in'.[52]

The intoxication of the past can make it a goal beyond price: du Maurier's hero risks health and even life for his excursions back to the 14th-century folk he comes to care for; 'the urge to see, to listen, to move amongst them was so intense' that it became obsessive.[53] 'To know that just by turning a few dials you can see and watch anything, anybody, anywhere, that has ever happened' makes a viewer 'feel like a god';[54] visiting the world of Ovid, Virgil and Catullus gives a traveller 'immense inner satisfaction'.[55] To go back to the late 17th century as his rich and well-born namesake at the court of

Charles II, John Dickson Carr's hero seems ready to sell his soul to the devil.[56]

Many science-fiction travellers seek the past not as a permanent home but simply as an occasional experience. A Wilson Tucker time warden 'always felt stimulated in the presence of the ancients; . . . they excited him and he was pleased to be living among them for however brief a span'.[57] Another fictional time traveller makes frequent brief trips back to see New York grow, the Hindenberg zeppelin burn, Lindbergh and the Wright brothers fly, Apollo 11, the assassination of Lincoln, Custer's last stand, the signing of the Declaration of Independence, even Creation itself.[58]

Even more exotic than the remote in space, time viewing and travel are often shown as profitable adjuncts to the tourist trade. Live television shows of Israelites crossing the Red Sea, the Fall of Rome, the French and American Revolutions would attract huge audiences.[59] One entrepreneur plans to sell visits to 'Daily Life in Ancient Rome, or Michelangelo sculpting the Pietà, or Napoleon leading the charge at Marengo', and entertain more venturesome travellers with Babylon, Helen of Troy in her bath, and Cleopatra's summit conference with Caesar.[60] Potential tourists to the past envisage picking up Attic pottery, finding Inca treasure, mining South African diamonds; others fancy a picnic in the Palaeozoic. Brian Aldiss's travellers, chancing to meet in the Devonian, find the package-holiday past already overcrowded: '"We're making our way up to the Jurassic. Been there?" "Sure. I hear it's getting more like a fair ground every year"'.[61]

Some glimpse history from the safety of the present, like Aldiss's 'chronograb' machine showing whole days in the lives of 'A Mother Dinosaur', of 'William the Conqueror's Wicked Nephew', and of 'A Citizen of Crazed Plague-Ridden Stuart London'; advertisements promise that 'YOU'LL LAUGH AS YOU LEARN . . . will give you history-sterics as it brings each era nearer'.[62] Another showman peddles 'Pawley's Peepholes on the Past', bringing a 'Quainte Olde 20th Century Expresse [to] See Living History in Comfort' in village England: 'Was Great Great Grandma as Good as she Made Out? See the Things Your Family History Never Told You!'[63]

Others visit but make no impact on the past. Du Maurier's protagonist sees, hears and smells as he walks through 14th-century Cornwall, but cannot be seen, cannot touch anything, cannot interfere in any way; he could only 'move about in their world unwatched, knowing that whatever happened I could do nothing to prevent it.'[64] Still others directly intervene in the past or take trophies from it. Tucker's travellers hearing a Lincoln speech in 1856 are thrilled to talk with Illinois townsfolk and with Lincoln himself; safari hunters bring home *Tyrannosaurus* heads as proof of their visit to the Mesozoic. Still others enjoy being famous historical characters; one of Poul Anderson's *Guardians of Time* delights in acting as King Cyrus of Persia.[65] Farmer's immortal expedition leader opts to stay behind in the Pleistocene, becoming in turn Abraham's father, Hercules and Quetzalcoatl.[66]

Explaining history. The desire to know how and why things happened is another compelling motive for revisiting the past in these tales. Whatever history's insights, we cannot observe what actually took place. If only we could see the past itself, and not merely its records and traces! 'If you could actually *see* what took place in the past, without having to infer it by these laborious and uncertain methods', Arthur C. Clarke's palaeontologist feels it would 'save you a lot of trouble'.[67] But what science-fiction historians mainly seek from the past is to find new facts and solve old dilemmas. One potential time traveller envisages 'all the treasure houses of history waiting to be opened, explored, catalogued'; he wants 'to stand on the city wall of Ur and watch the Euphrates flood . . . to know how *that* story got into Genesis'.[68] To 'see into the past and watch history unfolding itself, back to the dawn of time', revealing all the great secrets of antiquity, stimulates Clarke's venturers.[69] Farmer's are more explicit:

> Think of the historical mysteries and questions you could clear up! You could talk to John Wilkes Booth and find out if Secretary of War Stanton was really behind the Lincoln assassination. You might ferret out the identity of Jack the Ripper. . . . Get the true story on Pearl Harbor. See the face of the Man in the Iron Mask, if there ever was such a person. Interview Lucrezia Borgia and those who knew her and determine if she was the poisoning bitch most people think she was. Learn the identity of the assassins of the two little princes in the Tower.[70]

Helping to establish Marcus Antonius's exact birthdate, photographing Correggio's paintings in his studio, and recording 'the sonorous voice of Sophocles reading aloud from his own dramas' are another time traveller's special pleasures.[71]

Fred Hoyle's music historian seizes an opportunity to see classical Greece at first hand, for 'here was a chance to settle all the controversies and arguments about ancient music'.[72] The historian–captain's fascination with the past aggravates the crew's plight in Edmondson's *Ship that Sailed the Time Stream*, for he has 'done it on purpose. . . . He wants to go on back instead of getting us home'.[73] Historians who seek to overturn the conventional wisdom may be especially anxious to see what actually happened. An advocate of ancient Carthage in Asimov's *The Dead Past* desperately needs a chronoscope to disprove the slander, spread by their enemies, that the Carthaginians immolated little children as sacrificial victims.[74]

To see or revisit the past would do more than ascertain, confirm, or disprove historical facts; it could lend history a new dimension. 'Think of the understanding it would give us . . . to watch the culture of a world develop, to see its artists at work, its builders, its philosophers, its spiritual leaders!'[75] If historians 'could go back in time and see what happened and talk to people who were living then', one of Simak's characters conjectures, 'they would

understand it better and could write better histories'.[76] You could 'see
everything with your own eyes, . . . verify every fact, study every move,
every actor'. Observing, yet also knowing how events turn out, you would
be able to 'write history as no one ever did before, for you'll be writing as a
witness, yet with the perspective of a different period'. Ward Moore's
historian could thus 'write of the past with the detachment of the present and
the accuracy of an eyewitness knowing specifically what to look for'.[77]

Origins obsess some who seek the past. Darwin's *Origin of Species*
engendered a spate of fictional visits to the dawn of mankind. Other time
travellers have sought the beginnings of fire, of agriculture, of Indo-
European languages, of Alexander's conquests, of Columbus's voyages, of
Washington's contact with the Sons of Liberty, of the guillotine and the Eiffel
Tower. A visitor to an immortal world looks for the oldest inhabitant, for 'if
the first of them are still alive then they might know their origin! They would
know how it began!'[78]

Curiosity about the personal past excites dreams of returning to find out
what happened to us, our parents, our grandparents. To see some key
episode or figure from our own background could yield crucial insights.
Genealogists in Simak's *Catface* serve clients who wish to talk with or take
pictures of ancient forebears. Scientology urges converts to recall their foetal
experience, even their conception.[79] Incest is suggested as 'the force behind
the predisposition to look back'; an Aldiss character would 'like to see my old
man courting my mother and making love to her,'[80] and a Ward Moore time
traveller confesses a 'notion to court my grandmother and wind up as my
own grandfather'.[81]

Foreknowledge is a siren that lures some; they long for the past because it is
completed, even completed by themselves. Whereas the future is uncertain
because unpredictable, the past is safely mapped, its perils circumscribed, its
pleasures secure. To be in the past is like acting in a play whose plot we alone
know. To live again in the past indeed portends a kind of immortality, an
awareness that our present life is part of a long continuum of time. Coexisting
in past and present convinces du Maurier's protagonist that 'there was no
past, no present, no future. Everything living is part of the whole. We are all
bound, one to the other, through time and eternity . . . By moving about in
time death was destroyed.' His vivid experience of 14th-century Cornwall
proved that 'the past was living still, that we were all participants, all
witnesses', hence more truly ourselves.[82]

How to get there. Getting into the past is a task that science fiction and
other imaginative literature resolve in myriad ways. Science – or pseudo-
science – not only retrieves past sights and sounds, but returns us bodily to
previous times. 'We think the past is gone', says Finney's physicist, 'because
the present is all we can see. We can't see the past, back in the bends
and curves behind us. But it's there.' Extending Einstein's unified field

theory, 'a man ought somehow to be able to step out . . . and walk back to one of the bends behind us'. If Einstein is right 'the summer of 1894 *still exists*'.[83]

Drugs, dreams, knocks on the head, pacts with the devil, lightning bolts and thunder claps are common trans-temporal devices – some of them inadvertent. A flash of lightning on an operating still – a coil in a partially evacuated bell-jar – powers Edmondson's *Ship That Sailed the Time Stream*. Some devices restrict the period seen or visited: Edmondson's ship goes back just 1000 years at a jump; time in P. S. Miller's fantasy is coiled like a huge spring, permitting temporal transfer only in multiples of 60 million years.[84] But other time machines give viewers or travellers the epoch of their choice.

Relics of antiquity, often some small but brilliant or evocative artifact, trigger transitions from present to past in children's literature as well as science fiction. Swords and scabbards, votive axes, remnants of crosses and great grandmother's fan serve, like churingas among Australian aborigines, to bridge time. A fossilised sword arouses ancestral memories in a hero who 'recognises' it was long ago his own.[85] A Celtic sword hilt found off the coast of Maine enables Betty Levin's 20th-century children to travel back to Iron Age Ireland and 7th- to 10th-century Orkney, and participate in ancient folk life.[86]

Artifacts, houses and landscapes lead Penelope Lively's children into past worlds. A painted shield puts a 13-year-old in uncanny touch with the New Guinea tribe that had given it to her ethnologist great-grandfather.[87] In a Victorian house battered by time and use, 'actual, real things, like the sampler, and the initials on the table in her room', convince a young visitor that the past is real. She finds an old swing buried in the garden 'because it's been making noises to tell me it was', and decides that places, like clocks, 'go on and on, with things that have happened hidden in them, if you can find them'; perhaps places, too, can stop 'so that a moment goes on . . . for ever', letting you see into other people's time.[88]

Age-old relics in these tales hand down the past they contain to present-day beholders. Rubbing an Elizabethan silk bobbin boy against her cheeks 'to get the essence of the ancient thing', the heroine of Alison Uttley's *A Traveller in Time* found it 'smooth as ivory, as if generations of people had held it to their faces, and I suddenly felt a kinship with them, a communion through the small carved toy'.[89] Among keepsakes in an attic, an old man in a Ray Bradbury story could feel time breathing.

> It was indeed a great machine of Time, this attic, . . . if you touched prisms here, doorknobs there, plucked tassels, chimed crystals, swirled dust, punched trunk-hasps, and gusted the vox-humana of the old hearth-bellows until it puffed the soot of a thousand ancient fires into your eyes. . . . Each of the bureau drawers, slid forth, might contain aunts and cousins and grandmamas, ermined in dust.[90]

The marks of crucial past events may be so potent and enduring that they enter the consciousness of occupants centuries later. Thus at Uttley's Derbyshire house where Mary Queen of Scots tried a vain escape,

> the vibrant ether had held the thoughts of the perilous ruinous adventure, so that the walls of Thackers were quickened by them, the place itself alive with the memory of things once seen and heard. . . . The spoken words, the desperate knowledge of the queen's execution three hundred years before, had lain in some pocket of the ether and . . . was of such tragic import it pervaded my mind and became the most outstanding memory.[91]

To relive the past requires not just re-entry but wholehearted immersion. Empathetic association, profound and detailed knowledge, total familiarity with the chosen epoch are prerequisites for the time traveller, who must also avoid perplexing – or antagonising – people he meets in the past. The historian in Carr's *Devil in Velvet* is said to be the only man of the 1920s with sufficient knowledge of the minutiae of 17th-century life to carry off a return to it successfully.

Return to the past often requires prior habituation to historically realistic, if not authentic, environments. Finney's potential time travellers live for months in mock-ups reproducing the sights and sounds and smells of their destinations, wearing the clothes, eating the food, speaking the dialect of the time, so that when hypnotised back they will feel completely comfortable. Such preparation needs authentic stage sets. To accustom one expectant traveller to life 50 years back, a small town is stripped of modern anachronisms, and 'the bakery will be ready with string and white paper to wrap fresh-baked bread in. There'll be little water sprays in Gelardi's store to keep the fresh vegetables cool. The fire engine will be horse-drawn, . . . and the newspaper will begin turning out daily duplicates of those it published in 1926'. Fidelity to marks of wear is essential too: acid is used to stain every stone in a reproduction of Notre Dame, in order to prepare someone for a return to 1451 when the cathedral was already several centuries old.[92]

Rediscoveries and reconstructions

Beyond the realm of science fiction, how can we attain the felicities past landscapes hold in store for us? 'The past is not a peaceful landscape lying there behind me, a country in which I can stroll wherever I please, and which will gradually show me all its secret hills and dales', warns Simone de Beauvoir. 'As I was moving forward, so it was crumbling. Most of the wreckage that can still be seen is colourless, distorted, frozen: its meaning escapes me.'[93]

Except through imaginative reconstruction, the past is forever closed to us; tied to the here and now, we must make do with the present moment and with attenuated memories of a tiny span of time. The present is a prison from which we only dream of escaping.

Then is the past – that dear tapestry of the familiar, that storehouse of memory and monument, of artifacts and works of art, of the dreams of our progenitors – is it all irrevocably gone? Is there no way to recapture, re-experience, relive it? We reject so sad a denial and seek assurance that the past endures in recoverable form. Some agency, some mechanism, some faith will enable us to regain it, not merely as knowledge but as perception and feeling. The daily life of our grandparents, the rural sounds of yesteryear, the conversations of Rousseau, the deeds of the Founding Fathers, the creations of Michelangelo, the glory that was Greece can be experienced afresh. Somehow, sometime, a vehicle or a medium will carry us back. 'Not a thing in the past', suggested H. G. Wells, 'has not left its memories about us. Some day we may learn to gather in that forgotten gossamer, we may learn to weave its strands together again, until the whole past is restored to us.'[94]

Virginia Woolf poignantly yearned for the past: 'Is it not possible,' she wondered,

> that things we have felt with great intensity have an existence in-dependent of our minds; are in fact still in existence? And if so will it not be possible, in time, that some device will be invented by which we can tap them? I see it – the past – as an avenue lying behind; a long ribbon of scenes, emotions. . . . Instead of remembering here a scene and there a sound, I shall fit a plug into the wall; and listen in to the past . . . I feel that strong emotion must leave its trace; and it is only a question of discovering how we can get ourselves again attached to it, so that we shall be able to live our lives through from the start.[95]

Both our own experiences and those reported by others buttress faith that we can relive the past. 'We often live in the past', noted a chronicler, 'even . . . losing ourselves in some bygone period. We enter, in a particular way, into the lives of men and women of other times, seem to share their experiences and even their thoughts. We may hold imaginary conversations. We live simultaneously in the present and the past'.[96]

Renaissance and Enlightenment worthies, enthralled by classical Greece and Rome, spoke of their heroes as though they were actual contemporaries. Petrarch felt himself to be among the classical authors he read; 'It is with these men that I live at such times and not with the thievish company of today', as he 'told' Livy.[97] The ruins of ancient Rome stimulated the architect Filarete to 'see' the noble edifices as they had been in classical times.[98] Scholars wrapped themselves in the togas of Cicero and Lucretius to re-enact historic scenes. 'Ceaselessly occupied with Rome and Athens,' wrote Rousseau while reading Plutarch, 'living . . . with their great men, . . . I thought myself a Greek or a Roman.' The philosopher Holbach hunted and walked for days on end 'without forgetting the ever-captivating conversation of Horace, Vergil, Homer and all our noble friends of the Elysian fields.'[99] When Holbach wrote of wandering through fields and pastures with these ancient companions, he was not engaging in a pleasant conceit but reporting an emotional reality, whether he in fact visited antique scenes or instead joined admired ancients in present-day landscapes.

This mode of experiencing the past still carries conviction, if advertisements are any evidence: 'Newton, Cromwell, Byron, Milton, Tennyson, Pepys, Darwin: You ought to try living with them some time,' an American university tempts students to its extension programme in Cambridge.

> You ought to try it this summer. Sit under the same apple tree that gave Sir Isaac Newton a headache – and the world the theory of gravitation. Stroll through the courts, quads, and pathways that inspired Milton, Pepys, and Tennyson. And where Oliver Cromwell found his first following.[100]

An evocative variant of such communion is Bernard Levin's handshake game, tracing back, through forebears and mentors, figures from the past we and they in turn have met – a game which recently led scores of *The Times* correspondents to recall hand-to-hand contact with antiquity. Contemplating a chair inherited from an eminent great-grandfather, one wrote, 'I like to think that I am sitting in the seat of the mighty.'[101]

Another mode of experiencing the past is the rediscovery of history and prehistory. 'He who calls what has vanished back again into being,' wrote the historian Niebuhr, 'enjoys a bliss like that of creating.'[102] The thrill of historical rediscovery was keenly experienced by 18th- and 19th-century archaeologists and geologists who excavated Pompeii, stumbled on the secret of cuneiform, solved the mystery of the Rosetta Stone, and unearthed layers of geological history, extending knowledge back millions of years. Such explorations of the past enlarged the experience of history, with which the educated public felt a deep affinity. Although similar excitement still attends many archaeological discoveries, such historical passion is less characteristic of this century than of those before it.[103]

Travel is nowadays a regular highway to landscapes of yesteryear. Places that lag behind the modern mainstream, half-forgotten enclaves of bygone worlds, retain the flavour of earlier epochs. Americans visit such historylands so habitually that Stephen Spender taxes them with treating 'history as though it were geography, themselves as though they could step out of the present into the past of their choice'.[104]

Remnants of empire especially conserve the flavour of the past. Club life in Hong Kong or Singapore strikes the modern British observer as turn-of-the-century or earlier in dress, demeanour and standards of service. When 'an English nanny type' in Sydney brought him morning coffee on a silver salver, the journalist Peregrine Worsthorne marvelled that 'these delightful dodos, extinct in England, were still extant in the former colonies'. Australians seen on city streets seemed to him not 'like the British of today; more like the British looked two hundred years ago', or the 'gnarled codgers' in Victorian numbers of *Punch*.[105]

Encounters with ancient artifacts also make the past live again, in reality as in science fiction. In the landscape, physical relics can provide unmediated impressions of former times. Seeing history on the ground is a less self-conscious process than reading about it. Whereas texts require deliberate engagement with the past, tangible relics may come to us without conscious aim or effort. 'More open than the written record', in Lewis Mumford's words, 'buildings and monuments and public ways . . . leave an imprint upon the minds even of the ignorant or the indifferent.'[106] Ancient artifacts remain directly available to our senses, like Edwina's 'jagged piece of something from the Bronze Age' in L. P. Hartley's *The Collections*, 'the indescribable colour of age', which 'exhaled a whiff of antiquity'.[107]

American writer Helene Hanff is overwhelmed by London's historical flavour: 'I went through a door Shakespeare once went through, and into a pub he knew. We sat at a table against the back wall and I leaned my head back, against a wall Shakespeare's head once touched, and it was indescribable'.[108] The taste, the feel and the sight of historic objects etch them into memory as words alone can never do, and may also conjure up vivid images of their milieux. 'Picking up for one's self an arrowhead that was dropped centuries ago and has never been handled since', Hawthorne fancied he had received it 'directly from the hand of the red hunter', building up in his mind's eye 'the Indian village and its encircling forest, . . . recalling to life the painted chiefs and warriors, the squaws at their household toil, and the children sporting among the wigwams, while the little wind-rocked papoose swings from the branch of the tree.'[109]

The historian who surveys the locale of his work heightens its impact. Margery Perham's biography of Lord Lugard gains verisimilitude from her own Nigerian tour of duty; Bruce Catton obviously trod the Civil War battlegrounds he describes; Samuel Eliot Morison made a historical virtue of tracking Columbus's voyages by sail. For Gibbon, the actual visit to Rome, seeing 'each memorable spot where Romulus *stood,* or Tully spoke, or Caesar fell,' was crucial: 'On the 15th of October, 1764, as I sat musing amidst the ruins of the Capitol, while the barefooted friars were singing vespers in the Temple of Jupiter, . . . the idea of writing the decline and fall of the city first started to my mind.'[110] What enlivened the past for Gibbon was not simply the imperial ruins, but the echoes of the age-old pagan–Christian conflict symbolized by the friars in the temple.

Tangibility enhances the impact of historical fiction too. 'Does not our sense of that classic struggle between Holmes and Moriarty quicken', ask historians Middleton and Adair, 'if we have seen the Reichenbach Falls above the Englischer Hof?'[111] Recreated locales also can convey historical immediacy. Poussin made physical models of the Biblical, Greek and Roman scenes that inspired his Arcadian pictures; these painstaking reconstructions enabled the painter to see the past with his own eyes and feel it with his own hands.[112]

A past that lacks tangible relics is too abstract to be credible. Ruskin grumbled that because England has only 'a past, of which there are no vestiges; . . . the dead are dead to purpose. One cannot believe they ever were alive, or anything else than what they are now – names in school-books'. By contrast, in Italy 'at Verona we look out of Can Grande's window to his tomb', and feel that 'he might have been beside us last night'.[113] To be certain there was a past, we must see at least some of its traces. 'Like the very old, once-famous people whom everyone has thought dead for half a century but who turn out to be living in a furnished room somewhere, with half a dozen cats', places redolent of age prove, for Paul Zweig, 'that the past really existed once, that it wasn't made up by experts on the basis of archives'.[114]

The sense of coexistence with the present is another special quality of the tangible past. Something old, even something fabricated to seem old, can make us feel that the past lies before us, palpable and potent. Crusading to preserve Boston's Old South Church, a century ago, Wendell Phillips insisted that the Revolutionary heroes 'Adams and Warren and Otis are today bending over us, asking that the scene of their immortal labors shall not be desecrated or blotted from the sight of man'.[115] Disneyland's moving, speaking model of Abraham Lincoln helps us believe in our past by bringing it into the present, 'not be history *was* but history *is*', feels Ray Bradbury.[116]

Artifacts are simultaneously past and present; their historical connotations coincide with their modern roles. Landscape reinforces this temporal coexistence, commingling old with new. A flavour of antiquity permeates a row of houses, famous for architects and residents of various epochs, precisely because each house has endured through different times to make up the present ensemble. The mass of ancient artifacts on Dorset's hills gave Thomas Hardy the sense of all history together in layered proximity, notes Hillis Miller, 'seen equally close and equally far away, [and] seen at once, as from the perspective of eternity'.[117]

Yesterday's relics thus extend today's landscapes. The sheer endurance of buildings carries habits and values 'over beyond the living group, streaking with different strata of time the character of any single generation', in Mumford's words.[118] It is worth saving old houses which 'have stood and watched the processes of change', as a character in a novel asserts; 'you must keep the shells inside which such things happen, in case you forget about the things themselves'.[119]

Pictures and images of things past have increasingly helped to convey viewers back through time. Historical fidelity was essential, Victorian novelists and painters believed, for their audiences fully to experience the portrayed past. The mediaeval costumes in Stothard's *Monumental Effigies,* a 19th-century antiquarian masterpiece, were expressly designed to help viewers 'arrest the fleeting steps of Time' and 'live in other ages than our own'.[120]

Reconstructions, copies, surrogates and images play an increasingly important role in our awareness of and ability to conjure up the lineaments of past landscapes. 'Thanks to the cinema, the twentieth century and its inhabitants stand in a different relation to time from any previous age,' writes David Robinson. 'We can conjure up the past, moving and – for the past 50 years – talking just like life. People and places long dead are resurrected: Tolstoy and Potzdamer Platz, Stalin, the Crystal Palace, Steve Donoghue, Churchill, Auschwitz.'[121]

Historical reconstructions often persuade visitors they are in the past, as noted above, or that the past is very much alive in the present. US Park Service managers of prehistoric Indian ruins are told that they must 'convey the notion to the visitor that the ancients who lived there might come back this very night and renew . . . the grinding of corn, the cries of children, and the making of love and feasting' – though this last dictum ought not to be taken 'too literally'.[122]

Defects and drawbacks

Efforts to recapture or relive the past remain fictional and visionary for good reason: the achievement is impossible. The past has no doubt taken place; we ourselves stem from it. But the past we know or experience is not what actually happened; it is contingent on our own views, our own perspectives, our own present. Just as we are products of the past, so is the known past partly an artifact of our own. No perceiver, however immersed in the past, can divest himself of the assumptions and knowledge of his own time. Our hopes and fears, expertise and intentions continually shape the past we remember, study, revere and reconstruct. That past does depend on what actually occurred, but is never identical with it.

Neither the actual nor the experienced past can ever be known like the present. We can never replicate, let alone restore, the last detail of any event, any intention, of any time gone by. Living-history programmes bring back or recreate the past in no real sense, one historian cautions her perhaps over-sanguine clients. 'No recipe exists from which to concoct the thoughts, values, and emotions of people who lived in the past. Even having steeped ourselves in the literature of the period, worn its clothes and slept on its beds, we never shed [today's] perspectives and attitudes'.[123] Knowledge of the past as though it were the present comes only to those who, under hypnosis or in trance, mistake past for present.

Even those who re-enact history to convince others seldom persuade themselves that they are in the past. 'I live in the present; I just visit the past,' a costumed apprentice at Old Sturbridge Village told me in 1978; another had trouble remembering to be historical; a third felt that to get into a car after an 18th-century working day stripped her of temporal illusions.

The continual – and unavoidable – revision of texts and artifacts at once enriches and impoverishes the past. 'The most faithful historians, even if they do not alter or exaggerate the importance of matters to make them more readable, . . . almost always leave out the meaner and less striking circumstances,' so that the remainder is distorted.[124] Descartes' observation still holds true. We are now perhaps less prone to highlight the uniquely magnificent. But the past usually seems more vivid than it was, and than the present is, because we and our predecessors have selectively preserved and emulated its greatest monuments and achievements, and in so doing have inevitably neglected the rest.

A past thus made vivid conforms to our expectations. Perceptions formed by modern scenes require stimuli which an unadorned past can seldom supply. Habituated to a far wider range of artifacts and locales than our forebears, we would scarcely notice, let alone admire, the drab products and muted images of most previous epochs. And what was originally vivid time has mainly dulled and fragmented; ruins are impressive in their right, but most old remnants, left unrestored, would seem meagre or tawdry by contrast with the newly minted present.

What we know of the past also conflicts with how we feel it should be experienced. For example, pre-Gutenberg books conjure up images of splendidly illuminated manuscripts, although we know that such work was scarce and seen by far fewer than are nowadays bedazzled by the paperbacks on any modern railway stall. But the illuminated manuscript is admired today both as art and as an emblem of beleaguered culture, whereas the tawdry paperback is more apt to symbolise the debasement of culture and art alike.

History tells us that everyday life for most was hard and mean, but romance and nostalgia evoke a colourful, high-spirited world of castles, cathedrals and chivalry – as with the modern lass in *Doctor Who and the Time Warrior* who chides her mediaeval captors for their overly authentic reek of savagery: 'I know things were pretty scruffy in the middle ages, but really! You might leave the tourists a bit of glamour and illusion'.[125] Historic reconstructions and re-enactments often tidy and sweeten the past to conform with contemporary illusions. Tourist complaints about the dress, poor posture and lack of enthusiasm of those who 'animate' the soldiers at the reconstructed 18th century Fortress of Louisbourg, Nova Scotia, have forced park officials to warn visitors that the soldiers look scruffy because they *were* scruffy, 'reflecting the boredom and lack of personal pride so characteristic of garrison life' back then.[126]

Many have doubted that their own age could match the glamour or the achievements of previous eras. Roman monuments to ancient worthies could not be equalled in his own day, thought Giovanni Dondi in the 14th century, for want of requisite sculptors' skills and great men to memorialise.[127] Eighteenth-century English writers, since they could not hope to surpass Homer and Virgil, Milton and Shakespeare, felt constrained simply to emulate them.[128] The American National Trust promotes historic preservation with the slogan 'They don't build them like they used to. And they never will again' – reinforcing the message that today's architecture is innately inferior to yesterday's.

The predisposition to prefer past to present stems from two common but erroneous perceptions. One is the tendency to recall only what was best and assume it was characteristic. We overlook 'the large overbalance of worthlessness that has been swept away', as Wordsworth put it, and seize on great remains as typical.

> In our imaginative voyaging through the past, we are like those travellers through the jungle who are told where the grave mounds of giants from earlier days may be found. When we find the grave . . . we then assume that he was typical . . . rather than that he had been given such a mound in the first place and then remembered simply because he happened to have been a giant.

The second error is to view past and present as equivalent in length and hence in productive capacity, neglecting the fact, Wordsworth continues, 'that the present is in our estimation not more than a period of thirty years', whereas 'the past is a mighty accumulation of many such periods'. Thus every epoch tends to feel itself 'empty' in comparison with the past.[129]

If it were possible to return to the past, however, it might well prove disappointing. Asimov suggests that viewing or revisiting the past could become a crippling addiction for those neurotically attached to their childhoods, to long-lost parents, or to a Golden Age.[130] But the apparent charms of yesteryear would seldom commend themselves to its actual residents. 'If you told George Washington what you liked about his time', an aspiring time traveller is cautioned, 'you'd probably be naming everything he hated about it'.[131] A modern English heroine in ancient Rome is chagrined to realise that she may have to wait 1500 years for a cup of tea.[132] Horrified by the stark reality of the past, a visitor to the late Roman Empire realises that 'no restoration included all the dirt and the disease, the insults and altercations' of the past.[133] A film director in Laumer's *Great Time Machine Hoax* fears that 'practically everything in ancient history was too dirty for the public to look at.' Cold truth deflates great historic moments; William the Conquerer, a paunchy man of middle age in ill-fitting breeches of coarse brown cloth, a rust-speckled shirt of chain mail, and a moth-eaten fur cloak, is seen simply yawning and belching at the news from the Battle of Hastings.[134]

Great historic figures also disappoint Lafferty's time viewers, who are revolted by Aristotle's 'barbarous north-coast Greek' and Tristan and Isolde's bear-grease pomade, adjudge Voltaire's famous wit a 'cackle' and Nell Gwynn 'a completely tasteless morsel', and hearing Sappho talk for three days only of having her cat spayed, consider the future world fortunate 'that so few of her works have survived'. A time machine that takes him back just 20 years frustrates its inventor;[135] 'I saw enough of the Depression so I don't want to spend my old age watching people sell apples'.[136]

The enduring dream of reliving the past, even if impossible to fulfil, illuminates feelings about both past and present. Disenchantment with today impels us to 'yearn for the yesterdays, . . . never realizing,' a critic of time travel writes, 'that today, bitter or sweet, anxious or calm, is the only day for us. The dream of time is a traitor and we are all accomplices to the betrayal of ourselves'.[137]

In this dream, landscapes play a crucial role. 'The past is indestructible because human history gets incarcerated in physical things forming the scenes in which it was enacted,' writes Miller, interpreting Hardy's reactions to Dorset's immemorial hills. 'As long as these things continue to exist, even in the form of archeological debris, the history they embody can be resurrected in the retrospective eye.'[138] Hardy exemplifies what Lively terms 'the age old response to the amazing fact that the world we live in is much older than we are ourselves'. Aware 'that we only pass through the world, make a faint mark on it, and hand it on to someone else,' we envisage the landscape as 'a shrine to the past.'[139] It is because landscape seems to change far more slowly than memory, Drabble suggests, that 'we feel such profound and apparently disproportionate anguish when a loved landscape is altered out of recognition; we lose not only a place, but a part of ourselves, a continuity between the shifting phases of our life.'[140] Change in the beloved monuments of the outside world threatens to fragment the still more transient lives they have come to represent.

Notes

1 Drabble, M. 1979. *A Writer's Britain: Landscape in Literature,* 270–1. London: Thames and Hudson.

2 Lowenthal, D. 1978. Finding valued landscapes. *Prog. Human Geog.* **2**, 373–418.

3 Lowenthal, D. 1975. Past time, present place: landscape and memory. *Geog. Rev.* **65**, 1–36.

4 Hoskins, W. G. 1955. *The Making of the English Landscape,* 139, 299. London: Penguin.
See the discussion of Hoskins in D. W. Meinig 1979. Reading the landscape: an appreciation of W. G. Hoskins and J. B. Jackson. In *The Interpretation of Ordinary Landscapes: Geographical Essays,* D. W. Meinig (ed.), 195–244. New York: Oxford University Press.

5 Hoskins, W. G. 1963. *Rutland: A Shell Guide,* 7. London: Faber and Faber.

6 Forster, E. M. 1960. *The Longest Journey*, xiii. London: Oxford University Press.
7 Williams, R. 1973. *The Country and the City*. London: Chatto and Windus.
8 Bradbury, M. and M. Orsler. 1960. Department of amplification. *New Yorker*, 2 July, 58–62.
9 Lively, P. 1979. *Treasures of Time*, 80–1. London: Heinemann.
10 Lively, P. 1977. *The Road to Lichfield*, 57. London: Heinemann.
11 Grant, Mrs Donald B. 1974. quoted in R. Murray Schafer, Listening. *Sound Heritage* **3:4**, 10–17, ref. on p. 10. I discuss this in another context in: The audible past. In *The Canada Music Book* 11/12 (1975–76), 209–17.
12 Guy, B. S. 1978. Letter, *The Times*, 18 March;
    Coggan (1978) letter, *The Times*, 16 March.
13 Bolitho, R. 1978. Losing the scent of '78. *The Daily Telegraph*, 29 December, 10.
14 M. T. 1978. An American looks on London. *J. Lond. Soc.* No. 403, October, 7.
15 James, H. 1875. *A Passionate Pilgrim, and Other Tales*, 104. Boston: Osgood.
16 Tindall, G. 1980. *The Fields Beneath: the History of One London Village*, 230–1. London: Paladin.
17 Finney, J. 1970. *Time and Again*, 399. New York: Simon and Shuster.
18 Adams, H. 1918. *The Education of Henry Adams: an Autobiography*, 474–98. Boston: Houghton Mifflin.
19 Toffler, A. 1971. *Future Shock*, 353–4. London: Pan.
20 *Old Sturbridge Village: An Exploration of the Motivations and Experiences of Visitors and Potential Visitors* (New York: Fine, Travis and Levine, 1979), quoted in Darwin P. Kelsey 1980. *Reflections on the character and management of historical and tourist parks in the 1980s*. Keynote address, 3rd Annual Conference, Australian Historical and Tourist Parks Association.
21 Nin, A. 1973. *Seduction of the Minotaur*, 61–2. Chicago: Swallow.
22 Sayer, S. 1981. Wild landscape: Dartmoor – the influence and inspiration of its past. In *Our Past Before Us: Why Do We Save It?* D. Lowenthal and M. Binney (eds) 134 London: Temple Smith.
23 Henry Cleere, interview with the author, 5 June 1978.
24 Lively, P. 1973. Children and memory. *Horn Book Magazine* **49**, 400–7, ref. on 405.
25 Chateaubriand, F.-R. de 1802. *Le génie de christianisme*. Quoted in W. K. Ferguson 1948. *The Renaissance in Historical Thought: Five Centuries of Interpretation*, 121. Boston: Houghton-Mifflin.
26 Tom Stephenson, interview with the author, 29 August 1978.
27 Pendon Museum, Oxfordshire.
28 Bate, W. J. 1971. *The Burden of the Past and the English Poet*, 22. London: Chatto & Windus.
    On the primitivist taste for the primordial in the arts, see R. Rosenblum 1967. *Transformations in Late Eighteenth Century Art*, 140–60. Princeton: Princeton University Press.
    M. Greenhalgh 1978. *The Classical Tradition in Art*, 189–90. London: Duckworth.
29 Quoted in Blume, M. 1980. The film quest for B.C. *International Herald Tribune*, 21 August, 14.
30 Dresco, Jean-Pierre, cited in C. Corner 1977. Saving Montreux's belle époque heritage. *International Herald Tribune*, 1 November, 14.
31 Hoskins, W. G. 1973. *English Landscapes*, 6. London: BBC; idem 1963. *Provincial England: Essays in Social and Economic History*, 228. London: Macmillan.
32 Macaulay, R. 1967. *Pleasure of Ruins*, 113, 127. New York: Walker.
33 Newbolt, H. 1906. *The Old Country: a Romance*, 5. London: Smith, Elder.
34 Cottle, T. J. 1976. *Perceiving Time: a Psychological Investigation with Men and Women*, 52–6. New York: Wiley. The interviewees were 426 male and 101 female

17- to 21-year-old Navy recruits at a Michigan college. They were asked, 'How much would you pay right now to bring back' an hour, a day, or a year of their own past, and of any time before they were born, assuming they had 'a lot of money, more money than you can possibly use.'

35 My thanks to David Pringle, who made available the incomparable resources of the Science Fiction Foundation library at North East London Polytechnic, including subject-matter classifications permitting an exhaustive review of works dealing with viewing or travelling to the past. Peter Nicholls kindly sent me proofs from his encyclopedia (Nicholls, P. (ed.) 1979. *The Encyclopedia of Science Fiction*. London: Granada). The following articles were particularly useful: Adam and Eve, Alternate worlds, Origin of man, Reincarnation (Brian Stableford); Atlantis, Pastoral (David Pringle); History in S.F. (Tom Shippey); Mythology (Peter Nicholls); Time paradoxes, and Time travel (Malcolm J. Edwards). For the most notorious iron-age re-enactment, see Percival, J. 1980. *Living in the Past*. London: BBC.

36 Bester, A. 1958. Hobson's choice. In *Starburst*, 133–48, ref. on 148. New York: Signet.

37 Tucker, W. 1958. *The Lincoln Hunters*, 43. London: Hodder & Stoughton.

38 Simak, C. D. 1980. *Catface* (US title: *Mastodonia*), 54. London: Methuen/Magnum.

39 Farmer, P. J. 1975. *Time's Last Gift*, 79, 137. London: Granada/Panther.

40 Eklund, G. 1975. *Serving in Time*, 109. Don Mills, Ontario: Harlequin/Laser Books.

41 Simak, C. D. op. cit., 241.

42 Silverberg, R. 1979. *The Time Hoppers*, 156. Glasgow: Collins/Fontana.

43 McCarthy, M. 1949. *The Oasis*, 42–3. New York: Random House.

44 Finney, J. op. cit., 398.

45 Gerrold, D. 1973. *The Man Who Folded Himself*, 122. London: Faber & Faber.

46 Tucker, W. op. cit., 43.

47 Carson, R. 1957. *Pawn of Time: an Extravaganza*, 43. New York: Henry Holt.

48 Anderson, P. 1964. *Guardians of Time*, 68. London: Pan.

49 Du Maurier, D. 1970. *The House on the Strand*, 267. London: Pan.

50 Merwin, S., Jr 1960. *Faces of Time*, 33. London: Badger.

51 Anderson, P. 1977. The long remembering. In *Trips in Time*, Robert Silverberg (ed.), 78–92. Nashville and New York: Nelson.

52 Finney, J. op. cit., 218–19.

53 Du Maurier, D. op. cit., 241.

54 Sherred, T. L. 1947. E for effort. *Astounding Science Fiction* **30**(3), 119–62, ref. on 123.

55 Merwin, S. Jr op. cit., 7.

56 Carr, J. D. 1957. *The Devil in Velvet*. London: Penguin.

57 Tucker, W. op. cit., 112.

58 Gerrold, D. op. cit., 63–6.

59 Ballard, J. G. 1974. The greatest television show on earth. In *Low-flying Aircraft and Other Stories*, 148–54. London: Jonathan Cape. Sherred, T. L. op. cit.

60 Laumer, K. 1963. *The Great Time Machine Hoax*, 35. New York: Grosset & Dunlap/Ace.

61 Aldiss, B. 1969. *An Age*, 17. London: Sphere.

62 Aldiss, B. 1955. Not for an age. In his *Space, Time and Nathaniel*, 71–9, ref. on 76–8. London: Granada/Panther (1979).

63 Wyndham, J. 1956. Pawley's peepholes. In his *The Seeds of Time*, 96–120, ref. on 110. London: Penguin.

64 Du Maurier, D. op. cit., 40.

65 Anderson, P. op. cit., Ch. 4.
66 Farmer, P. J. op. cit., 167, 171.
67 Clarke, A. C. 1950. Time's arrow. In his *Reach for Tomorrow*, 132–48, ref. on 139. London: Corgi.
68 Tucker, W. 1972. *The Year of the Quiet Sun*, 84. London: Arrow.
69 Clarke, A. C. op. cit., 143.
70 Farmer, P. J. 1974. *To Your Scattered Bodies Go*, 40. London: Granada/Panther.
71 Tucker, W. op. cit., 112.
72 Hoyle, F. 1968. *October the First is Too Late*, 96. London: Penguin.
73 Edmondson, G. C. 1965. *The Ship that Sailed the Time Stream*, 92. New York: Ace.
74 Asimov, I. 1956. The dead past. In his *Earth is Room Enough*, 9–50, 25. London: Granada/Panther (1960).
75 Merwin, S. Jr. op. cit., 115.
76 Simak, C. D. op. cit., 54.
77 Moore, W. 1955. *Bring the Jubilee*, 159–60, 169. London: New English Library (1977).
78 Lafferty, R. A. 1966. Nine hundred grandmothers. In (1970) *Nine Hundred Grandmothers*, 10. New York: Ace.
See Harrison, H. 1965. Famous first words. *Fantasy and Science Fiction* **28**(1), 67–71.
79 Hubbard, L. R. 1950. *Dianetics: the Modern Science of Mental Health*, 237. Los Angeles: Church of Scientology (1979).
80 Aldiss, B. *An Age*, op. cit., 29.
81 Moore, W. op. cit., 164.
82 Du Maurier, D. op. cit., 169–70.
83 Finney, J. op. cit., 52, 63.
84 Miller, P. S. 1937. The sands of time. In *Great Science Fiction Stories*, C. T. Smith (ed.), 231–66. New York: Dell (1964).
85 Ashton, F. 1946. *The Breaking of the Seals*, 26. London: Andrew Dakers.
86 Levin, B. 1973, 1975, 1976. *The Sword of Culann; A Griffon's Nest; The Forespoken*. New York: Macmillan.
87 Lively, P. 1977. *The House in Norham Gardens*, 67–8. London: Pan.
88 idem 1978. *A Stitch in Time*, 48, 104, 113, 139. London: Pan.
89 Uttley, A. 1939. *A Traveller in Time*, 50. London: Puffin (1978).
90 Bradbury, R. 1963. A scent of sarsaparilla. In his *The Day It Rained Forever*, 192–8, ref. 196–7. London: Penguin.
91 Uttley, A. op. cit., 106.
92 Finney, J. op. cit., 65, 48.
93 Beauvoir, S. de 1978. *Old Age*, 407. London: Penguin.
94 Wells, H. G. 1929. *The Dream*, 236. London: Collins.
95 Woolf, V. 1976. *Moments of Being*, 67. Sussex: The University Press.
96 Woodward, W. A. 1934. *The Countryman's Jewel: Days in the Life of a Sixteenth Century Squire*, xi–xii. London: Chapman & Hall.
97 Petrarch, F. 'Familiar letters'. In Burke, P. 1969. *The Renaissance Sense of the Past*, 22. London: Edward Arnold.
98 Filarete, A. *Treatise on Architecture*, quoted in Erwin Panofsky 1970. *Renaissance and Renascences in Western Art*, 20. London: Granada/Paladin.
99 Rousseau, J.-J. Confessions (1781) and Paul Henri Dietrich d'Holbach (1746), quoted in Gay, P. 1966. *The Enlightenment: an Interpretation, I: The Rise of Modern Paganism*, 46, 54. New York: Knopf.
100 UCLA Extension advertisement 1980. *New York Review of Books*, 22 January, 20.
101 Levin, B. 1980. How to shake hands with a legend. *The Times*, 5 March, 16.
Higham, M. 1980. Letter. *The Times*, 26 March, 17.

102 Niebuhr, B. G. 1828. *The History of Rome*. Quoted, with approval, by Lyell, C. 1830. *The Principles of Geology*, vol. 1, 74. London: Murray.
Rudwick, M. 1979. Transposed concepts for the human sciences in the early work of Charles Lyell. In *Images of the Earth: Essays in the History of the Environmental Sciences*, L. J. Jordanova and R. S. Porter (eds), 67–83, ref. on 68. London: British Society of the History of Science.

103 Plumb, J. H. 1973. *The Death of the Past*, 111. London: Penguin.
See also Carter, L. *Imaginary Worlds*, 51. New York: Ballantine.

104 Spender, S. 1974. *Love–Hate Relations: a Study of Anglo-American Sensibilities*, 121. London: Hamish Hamilton.

105 Worsthorne, P. 1979. Home thoughts from Down Under. *Sunday Telegraph*, 25 February, 8–9.

106 Mumford, L. 1938. *Culture of Cities*, 4. London: Secker & Warburg.

107 Hartley, L. P. 1972. *The Collections*, 53. London: Hamish Hamilton.

108 Hanff, H. 1976. *84 Charing Cross Road*, 118. London: Futura.

109 Hawthorne, N. 1842. *Mosses from an Old Manse. Works*, Riverside edn, vol. II, 20. Boston: Houghton Mifflin (1882).

110 Gibbon, E. 1796. *Autobiography*, 84–5. London: Routledge & Kegan Paul (1970).

111 Middleton, A. P. and D. Adair 1969. The case of the men who weren't there: problems of local pride. In *The Historian as Detective: Essays on Evidence*, R. W. Winks (ed.), 142–77, ref. on 173–4. New York: Harper and Row.

112 Praz, M. 1969. *On Neoclassicism*, 28–9. Evanston, Ill.: Northwestern University Press.

113 Ruskin, J. 1886. *Modern Painters*, vol. 4, 4–5. New York: Wiley.

114 Zweig, P. 1977–8. Paris and Brighton Beach. *Am. Scholar* **47**, 501–13, ref. on 512.

115 Phillips, W. 1965. In *The Old South Meeting House*, quoted in Hosmer, C. B. Jr *Presence of the Past: a History of the Preservation Movement in the United States before Williamsburg*, 104. New York: Putnam's Sons.

116 Bradbury, R. 1965. The machine-tooled happyland – Disneyland. *Holiday* **38**(4), October, 100–4.

117 Miller, J. H. 1972. History as repetition in Hardy's poetry: the example of 'Wessex Heights'. In *Victorian Poetry*, M. Bradbury and D. Palmer (eds), 222–53, ref. on 227–8. London: Edward Arnold.

118 Mumford, L. op. cit., 4.

119 Lively, P. *House in Norham Gardens*, op. cit., 12.

120 Stothard, C. A. 1811–33. *The Monumental Effigies of Great Britain*, quoted in Strong, R. 1978. *And When Did You Last See Your Father? The Victorian Painter and British History*, 55. London: Thames and Hudson.

121 Robinson, D. 1979. The film immutable against life's changes. *The Times*, 7 December, 11.

122 Tilden, F. 1957. *Interpreting our Heritage: Principles and Practices For Visitor Services in Parks, Museums and Historic Places*, 69. Chapel Hill: University of North Carolina Press.

123 Sherfy, M. 1976. The craft of history. *In Touch* (interpreters' information exchange), no. 13, May 4–7.

124 Descartes, R. 1637. Discourse on the method of rightly directing one's reason and of seeing truth in the sciences. In Anscombe, E. and P. Keach 1969. *Descartes: Philosophical Works* (translated and edited), 11. London: Nelson.

125 Dicks, T. 1978. *Doctor Who and the Time Warrior*, 64. London: W. H. Allen/Target.

126 Proudfoot, D. 1976. How Louisbourg restored looks today. *Can. Geog. J.* **93**(1), 28–33, ref. on 30.
See also Fortier, J. 1978. Thoughts on the re-creation and interpretation of

historical environments. *Third Conference Proceedings, International Congress of Maritime Museums,* 251–62. Mystic, Conn.

127 Dondi, Giovanni, cited in Panofsky, E. op. cit., 208.

128 Bate, W. J. op. cit.

129 Wordsworth, W. 1809. Answer to Mathetes. In *The Friend,* S. T. Coleridge (ed.), no. 17, 14 December. Reprinted in Rooke, B. E. (ed.) 1969. *Collected Works of Samuel Taylor Coleridge,* vol. 4: II, 231. London: Routledge and Kegan Paul. The indented quotation is paraphrased from Wordsworth by Bate, W.J. op. cit., 70.

130 Asimov, I. The dead past, op. cit.

131 Bester, A. Hobson's choice, op. cit., 146.

132 Merwin, S., op. cit., 13.

133 Sprague de Camp, L. 1941. *Lest Darkness Fall,* 12. New York: Pyramid (1963).

134 Laumer, K. op. cit., 36–7, 47.

135 Lafferty, R. A. Through other eyes. In *Nine Hundred Grandmothers,* op. cit., 282–96, refs on 282–4.

136 Gross, M. 1952. The good provider. In Conklin, G. 1955. *Science Fiction Adventures in Dimension,* 167–71, ref. on 170. London: Grayson & Grayson.

137 Bester, A. Hobson's choice, op. cit., 148.

138 Miller, J. H. History as repetition, op. cit., 246.

139 Lively, P. 1978. Children and the art of memory: part II. *Horn Book Magazine* **54,** 197–203, ref. on 202.

140 Drabble, M. op. cit., 240–1.
     See also Goldstein, L. 1973. The Auburn syndrome: change and loss in 'The Deserted Village' and Wordsworth's 'Grasmere'. *ELH* **40,** 352–71.

6 *Landscape aesthetics: a synthesis and critique*

JOHN V. PUNTER

Questions of landscape aesthetics have rarely been approached directly by those concerned with environmental values. Rather, researchers have preferred to discuss questions of preference, taste, perception, interpretation, evaluation, management and modification as more tangible surrogates or indicators of aesthetic experience. Explicit discussion of landscape aesthetics largely disappeared with the eclipse of Romanticism, though it has been kept alive throughout the 20th century by the work of a few, largely isolated individuals: notably, Vaughan Cornish in his writings on aesthetic geography, the work of Gordon Cullen and others on townscapes, and a small cadre of human geographers who have contributed extensively to the journal *Landscape*.[1] Nevertheless, there has been a recent revival of interest in the search for correctives to positivist approaches to landscape and spatial experience, and with the preoccupation of more affluent, mobile and leisured members of the community (and thereby institutional and governmental policy-makers) with questions of environmental amenity. Evidence of this revival can be seen in a diverse body of new work which tackles questions of aesthetic experience from radically different philosophical and methodological perspectives, which have parallels, but apparently little direct contact, with developments in aesthetic theory at large.

This chapter briefly reviews some of the newer literature on landscape aesthetics, categorises it according to basic research paradigms, and examines its philosophical orientation. Instead of the traditional atheoretical eclecticism that characterises writing in this field, it advocates the development of a materialist perspective as a means of redefining the preoccupations of landscape aesthetics itself, as a vehicle for a critical synthesis of current developments, and as a means of identifying future research questions.

Landscape aesthetics: a broader field of enquiry

As a point of departure, it is necessary to examine the conceptions of landscape aesthetics that underlie this diffuse body of literature. The lack of theoretical base and the gap between the philosophy of aesthetics in general and landscape aesthetics in particular have been demonstrated elsewhere, as has the fundamental split between work on landscape experience and landscape interpretation.[2] The latter split is especially significant, and while there are authors whose work transcends both categories, the gulf is enforced by disciplinary boundaries. Generally, landscape aesthetics is considered as separate from both experience and interpretation, as somewhat peripheral and inconsequential: the preserve of the *dilettante* concerned with 'high level' esoteric experience. This particular attitude can be traced to the linguistic roots of the word 'aesthetics' itself.[3] The term only appeared in English in the 19th century and was first borrowed from German in a Latin form (*aesthetica*) to embrace the study of beauty as phenomenal perfection apprehended through the senses. By contrast, the Greek root of aesthetics contains a broader conception, and is not preoccupied with beauty *per se* but with the conditions of sensuous perception. It is this latter meaning which has been largely lost. Subjective 'sense activity' has become emphasised as the basis of art and beauty, and social or cultural interpretations have been played down in favour of philosophical treatises on 'the beautiful' and the 'principles of good taste'.

The contrast between the two meanings is clearer in the relevant antonyms 'unaesthetic' relating to the Latin root, and 'anaesthesia' to the Greek. It is a return to the latter and broader Kantian conception of aesthetics which is being sought in this chapter, the purpose being to integrate aesthetic with socio-cultural interpretations while recognising that the conventional aesthetic affirms a complex range of human meanings and values, such as imagination, identity and creativity.

A broader conception of 'landscape' is also necessary if the field of land-scape aesthetics is to contribute significantly to the understanding of valued environments. In examining the usage of the term 'landscape', a similar conventional displacement and subsequent marginality emerges. The origin of the word landscape as a technical term within (Dutch) landscape painting is significant, and its limitation to beautiful natural scenery decisive.[4] A second, later meaning of landscape as a view or prospect is similarly revealing, and the common 17th-century use of the term was to convey 'pleasing prospects', the objects of self-conscious contemplation.[5]

Here, then, is a significant link with the conventional aesthetic. Landscape excludes 'figures', activities and social setting, and emphasises rural ('natural') beauty. Aesthetic plays down the social and cultural in favour of subjective sense activity and similar notions of beauty. So landscape aesthetics concentrates on the 'pleasing prospects' – the visual backdrops – divorcing man and social process from the scene, struggling with the meaning of the relative beauty of the visual world, disclosing interpretational meanings, but largely divorcing these from conceptions of landscape experience. This is a mutually reinforcing limitation, and attempts to cir-cumvent it are clearly apparent in the body of literature that characterises the broadly conceived field of landscape aesthetics.

In reviewing contemporary developments in landscape aesthetics, it is often impossible to categorise neatly the variety and breadth of contributions, but three interdisciplinary perspectives or paradigms are immediately apparent. They embrace the main research thrusts of individual disciplines – architecture, landscape architecture, urban design, planning, geography, environmental psychology, sociology and even art and literary criticism – and thus are convenient vehicles for the purposes of review. The three paradigms might be entitled landscape perception, landscape interpretation, and landscape (visual) quality. Each has a clear focus of interest, though within each paradigm are an infinite variety of philosophical and methodological perspectives. Broadly, the *perception paradigm* embraces the mechanics of how we perceive landscape and the links between vision, perception, comprehension, preference and action. The *interpretation paradigm* focuses almost exclusively on the meanings imputed to landscape, and particularly its social and cultural content. Finally, the *landscape quality paradigm* focuses upon visual quality and qualities (formalism), largely excluding both the mechanics of perception and the question of meanings. These three paradigms in fact embrace almost all work of consequence on landscape form and experience, with those that it excludes being related to one of three paradigms by extension.

Research paradigms in landscape aesthetics

Landscape perception. The roots of the perception paradigm in psychology, and its subsequent adoption by the social and design sciences as a central element in the development of a behavioural perspective, are well known. The concern was to understand the response of the (predominantly visual) senses of an observer to external stimuli (primarily from the physical environment), to understand how things were 'seen', to learn how information was gathered, interpreted and understood, and to find out how this in turn affected attitudes, preferences and behaviour. The adoption of the perception paradigm by such a large body of researchers in such a disparate group of disciplines naturally diffused and corrupted the original meaning of the term 'perception' as propounded by psychologists.[6] The most significant source of confusion remained the identification of 'elements' or of 'stages' in the perception process. Rapoport made a useful distinction between three relatively distinct subprocesses within the perception process at large – perception itself, cognition, and evaluation (or preference). He limited perception to 'direct sensory experience', cognition to 'the way in which people understand, structure and learn', and evaluation to 'the perception of environmental quality, and hence preference'.[7] This tripartite division might be extended by separating vision from perception,[8] and separating evaluation from action (actual landscape modification, relocation or movement).

Armed with this five-fold classification of subphases of perception, albeit accepting that the classification has only limited theoretical support,[9] a number of significant contributions to landscape aesthetics within the perception paradigm may be identified. In particular, much is not understood about *vision* and the way in which optical laws and the physiological limitations of the eye structure landscape experience.[10] The nature of visual fields, focusing behaviour, and processing of successive retinal images are comparatively well understood, and interesting applications of these ideas have been made in both the design of urban space – differentiating various scales and degrees of enclosure – and the design of highways – for road safety and aesthetic experience.[11] These optical laws are of fundamental importance to landscape experience and have become an accepted part of design methodology.

With regard to *perception*, two recent applications of information theory have proved useful to an understanding of landscape experience. Preoccupation with questions of sensory deprivation and overload, or alternatively, preferences for complexity and ambiguity, and the search for optimal rates for perception have initiated the development of a whole field of psychological aesthetics,[12] which now provides an umbrella for a range of more intuitively based work in architecture and design. While the simple hypotheses currently formulated fail to do justice to the complexities of aesthetic judgement and landscape experience, they supply useful first approximations of the nature of aesthetic attitudes and interests. An indication of this may be found in the frequent cross-disciplinary references to such polarities as visual complexity *vs.* simplicity, visual chaos *vs.* order, ambiguity *vs.* clarity, commonplace *vs.* unique, and arousal *vs.* habituation.[13] The second relevant application of information theory to have appeared recently is the development of the *Gestalt* laws of form perception in an urban context.[14] These 'laws' are a series of conditional tendencies in perceptual organisation that hinge upon the law of *Pragnanz*, which itself indicates that the visual system seeks to integrate separate visual stimuli in a meaningful whole, maximising redundancy and simplicity of form. These conditional tendencies have been applied to the perception of built forms and have been used to explain the way that individuals conceptualise urban space.[15]

The conceptualisation of space is more a concern at the *cognition* phase of perception which stresses knowledge of, and meanings given to, the landscape. In general, environmental perception research has stressed knowledge of the environment and the cognitive schemata (urban images, mental maps, orientation, demarcation and symbolic devices) by which the chaotic world is transformed into useful knowledge.[16] This narrow, but well trodden, route to questions of cognition contrasts markedly with the lack of research on questions of meaning. Other than to differentiate between the connotative and denotative aspects of meaning,[17] environmental psychology has failed to develop adequate conceptions, partly because of the necessity to control the number of variables under consideration in laboratory situations. Preliminary experimentation suggests that the whole process of perception is fundamentally, if not overridingly, influenced by meaning – a conclusion confirmed by the few explicit studies of meanings in the urban landscape conducted so far.[18]

Research on *evaluation* has focused on a variety of topics within the broad spectrum of environmental quality assessment. Much has been learned about the relative importance and nature of the components of landscape quality,[19] but there has been little examination of these factors *per se* as elements of aesthetic experience. Clear differences emerge between professional and lay tastes,[20] but these are reported rather than their meaning examined. The most profound and controversial contribution to evaluation has come from attempts to apply the findings of neurophysiology and neuropsychology to questions of landscape preference. This work embraces the perception, cognition and evaluation subprocesses but its basic aim is to demonstrate that landscape preferences are formulated by the capacities and tendencies of the brain itself, and to use this hypothesis as a basis for a pyschological theory of aesthetics.[21] Certainly this research extends the more simplistic notions of experimental aesthetics and considers the infinite variety of experiences that can arise from the various sympathetic opposites in the brain, while arguing that, over and above this variety and individuality, there are large areas of consensus common to groups and cultures, and even, at the archetypal level, to human consciousness itself.

Action as the final stage in the perception (behavioural) process might seem only very indirectly related to questions of landscape aesthetics. However, through the extensive work of ethologists emphasising the non-learned and genetic aspects of animal and human behaviour, and the development of conceptualisations of home range, territory, jurisdiction, personal and defensible space, behavioural sinks and ritualised behaviour, a number of interesting observations have emerged on how people react to particular physical and social landscapes.[22] There have even been attempts to extend ethological interpretations to landscape experience itself via 'habitat theory',[23] but the greatest interest lies in psychological interpretations of the 'personalisation' of landscape as a direct expression of peoples' environmental preferences, tastes, and conceptions.[24]

From the perspective of those interested in the everyday experience of the urban landscape one might be tempted to despair at the current state of experimental environmental psychology with its complex, often apparently ridiculous, methodological trapppings.[25] Yet the perception paradigm has produced valuable insights into the nature of the visual system and the bases of perception. If it has faltered on questions of meaning and evaluation, that is only to be expected, for the complexities of these phenomena do not yield themselves easily to positivist approaches of experimental psychology or information theory. While research methodologies are attuned to the discovery of simple, concrete, use or instrument meanings, questions of emotional, sign or symbolic meanings have not entered the paradigm except as a general category of connotative meanings. Yet the foregoing provides sufficient evidence to demonstrate the very considerable contribution that the perception paradigm has made to our understanding of how people apprehend, understand, appreciate and evaluate their landscapes. Equally, it is clear that many of our conceptions are naïve, bear little relationship to actual perceptual or cognitive processes, and do absolutely no justice to the meaning of landscape experience or aesthetic experience as a whole. For these reasons, one can gain most out of this paradigm by interpreting its findings in the context of ideas generated in other paradigms.

Landscape interpretation. The landscape interpretation paradigm subsumes a wide range of work in archaeology, architecture and architectural history, history, anthropology, geography, planning and design. Inevitably, the focus and purpose of interpretation varies from discipline to discipline and from study to study, since the meanings sought differ according to the orientations of the researcher and his disciplinary training and since, in any case, all landscapes admit a multiplicity of meanings. We have already noted the almost wholesale neglect of the question of meanings in the perception paradigm. Within the interpretation paradigm, the same tendency is evident, with meaning being largely defined by disciplinary tradition. Architects tend to be preoccupied with the functional and symbolic meanings of architectural style; anthropologists and geographers with the landscape as an expression of social, economic and cultural life; and the social historian with the landscape as a composite expression of past social values and sentiments. In the wider view, however, all are approaches to the same central issue of the landscape as a reflection of human values and life-styles. As Tuan[26] suggests:

> We seek meaning in the landscape because it is the repository of human striving . . . Meaning implies two things. One is order or harmony. We find meaning when we can discern order or harmony in the chaotic world of facts and remove the irritation and insecurity that chaos generates. Meaning also implies significance: a phenomenon has meaning because it is a sign to something beyond itself, to its own past and future and to other objects. The significant object or event has the seeming capacity to condense the diverse strains of the universe into a thing within human reach.

Thus the comprehension of meaning involves the search for order and the search for significance. Clearly each discipline has its own conception of both order and significance but ultimately the goal is the same – understanding what the landscape means and has meant to various people.

The literature on landscape interpretation can be divided into several categories – cultural geography, local and urban history, architecture and semiotics. The work of cultural geographers on landscape demonstrates several different orientations, from large scale synthetic-descriptive works on regional character, to historical studies of relict landscapes, artifacts or landscape features. Nevertheless, the aesthetic dimension has always posed problems for geographers, who have treated it with a certain amount of superficiality, mysticism or highly subjective commentary.[27] Extensive use has been made of anthropological material on primitive societies[28] and of literary and art criticism,[29] and a close relationship has been developed between landscape interpretation and cultural history at large. Those writers who have successfully understood the visual appeal and sociocultural meanings of particular landscapes have themselves faced the stumbling block of an appropriate social perspective and philosophy for their interpretations.[30]

The recent growth of humanistic geography has re-established a better understanding of man as the goal of landscape studies and corrected the tendency to see landscape as an end in itself. Significantly, the term 'landscape' has been largely supplanted by the existentialist concept of *place* which re-emphasises physical setting (landscape), people and patterns of behaviour (social landscape), and meanings given to and derived from these settings, all as an indissoluble tripartite complex.[31] This refocuses the task of interpretation as the elucidation of meaning from physical and social settings and allows the adoption of a direct experiential perspective on landscape studies that can draw on, but is not bounded by, perception studies.

Within local and urban history, the emphasis has been primarily upon the 'morphogenesis' of the landscape, using it as an historical record or 'palimpsest',[32] but more interest has been focused on the processes involved than on the contemporary everyday significance of historic elements or interpretations.[33] Work on historic preservation suggests that visible landscape history – personal, regional and national – provides a rich source of cultural meanings that may in themselves be major components of preference and even emotional stability.[34] Architectural historians have been more productive in developing contributions to landscape aesthetics at large, having abandoned their preoccupation with the historic progression of styles of elite structures in favour of examining the building as a whole within its historical-cultural context.[35] The view of architecture as a mirror of society has emerged. Analysis remains largely at the level of *Zeitgeists*, 'tastes', or loosely defined 'fashions',[36] but issues of symbolism and analysis are being more thoroughly explored.[37] Questions of the relationship between cultural interpretation and aesthetic judgement continue to be discussed,[38] and these may be seen as symptomatic of both professional self-conceptions and of wider debates about the nature of the aesthetic.

Of more immediate relevance has been the application of semiology to architectural theory and criticism. Semiology – the science of signs – provides various theoretical perspectives, largely derived from linguistics, of obvious potential relevance to the study of the production and expressive meaning of architecture and the built environment. It offers a more rigorous approach to the question of how buildings communicate meanings, and how the landscape may be 'read', providing deeper analytical approaches to questions of iconography and symbolism, analogy and metaphor, denotation and connotation. As yet there is no consensus on the most appropriate theoretical frameworks for such analysis and little attempt to develop complementary findings.[39] The consequent conflicts over terminology and the multiplication of jargon make replication and synthesis difficult, if not impossible. Other problematic characteristics of architectural semiology also threaten to undermine its potential contribution to understanding landscape experience, and have led many writers to dismiss its utility altogether.[40] These characteristics include the limitations of a structuralist perspective at large, the focus upon meanings for architects (rather than the general public), the preoccupation with a relatively few architectural masterpieces for analysis, and the diffusion of interest into the generation of architectural form and semiology of the design process.

Each of the various approaches, however, can yield further insights into the interpretation of the meaning of the urban landscape, whether it is the study of simple analogies between architectural and non-architectural forms (e.g. Sydney Opera House and intersecting sea shells), the more complex relating of architectural signifiers (such as forms, spaces, surfaces, volumes, rhythms and textures) to signifieds (functional, iconographic, aesthetic associations and meanings), or the much more ambitious attempts to understanding the dialectic between the significative forms and codes of interpretation that produce the 'expressiveness' of buildings.[41] Despite the theoretical and practical difficulties, semiology provides an alternative starting point for an investigation of meanings in the landscape and a set of frameworks that can embrace and rework other traditions of interpretation (e.g. ethology and information theory), as well as a means of deepening our understanding of the cognitive processes of perception. Like humanistic geography, meanings for man can be placed at the centre of concern, avoiding both the general tendency towards abstracting the physical setting as the main concern and the trap of treating the description and analysis of landscape as an end in itself.

Yet few writers have broached the fundamental philosophical questions about interpretation that have been so prevalent in writings on aesthetics in general. Only in recent years have scholars like Jackson, Tuan, Gombrich and Williams begun to explore such issues as the relative emphasis that should be given to spiritual *versus* material, artistic *versus* utilitarian, and formal *versus* moral considerations in landscape interpretation. Nevertheless, there is a growing awareness that room exists for more rigorous and philosophically consistent interpretations of existing landscapes, and that such contrasting interpretations can serve to illuminate one another in the same way as the contrasting disciplinary perspectives of geography, history and architecture can complement each other in furthering an understanding of the forces shaping the urban landscape.

Landscape/visual quality. The third paradigm is in many ways the weakest in terms of substantive research. Paradoxically, it continues to exert an alarmingly strong influence on conceptions of how one experiences landscape, an influence which has been particularly significant in architecture, urban design and town planning. The phrase 'visual quality' is not entirely satisfactory, although preoccupation with a narrow conception of visual qualities does characterise approaches within this tradition. One might also describe it as an aesthete's approach in its preoccupation with beauty and the formal or artistic qualities in building and the landscape, and its emphasis upon subjective 'sense activity' as the basis of art and beauty. Clearly this philosophical stance is shared by many other researchers whose work could have been discussed in other paradigms, but this paradigm is primarily concerned with the visual, formal qualities of the landscape and only secondarily with the associated emotional and aesthetic effects. Especially important is the neglect of both the questions of meaning (that are so significant in the landscape interpretation paradigm) and the processes of perception (emotional and aesthetic effects of formal visual qualities are imputed but not analysed). These omissions suggest that the paradigm exists in its own right.

Perhaps the only substantive literature falling within this section is that of the 'townscape school'. At its best, this school of thought extends into questions of perceptual process, social use and the attachment of meanings,[42] but at its worst becomes 'the architectural equivalent of the girlie magazine'.[43] Many prominent landscape researchers have applauded and extended Cullen's preoccupation with human needs within the visual environment,[44] but critics have decried the 'disembodied eye' approach of later adherents to the school, whereby signification, cultural meanings and political implications of townscape policies are ignored in favour of a detailed, but naïve, formalism.[45] The same naïve formalism characterises attempts to develop notation systems to describe landscape experience,[46] and to develop evaluation procedures,[47] although, increasingly, professional perceptions are coming to be used as the 'arbiters of excellence'. Again, critics have emphasised the complete absence of aesthetic theory underpinning these research efforts,[48] though this absence is only more obvious in this paradigm by virtue of the explicit measurement criteria adopted.

Of the three paradigms, the visual quality paradigm yields the least material that fits exclusively within its bounds, and yet the focus on qualitative judgements of purely visual factors, to the detriment of discussion of meanings, emotional or psychological effects, remains pervasive within the study of landscape aesthetics. Take, for example, architectural or design studies of visual quality. While these tend to touch on questions of meaning and emotional effect, most remain preoccupied with questions of solid and void, space and structure, scale and proportion, rhythmical texture, light and colour, order and diversity.[49] These are most frequently discussed in the context of the great set pieces of architectural history and rarely that of the everyday landscape. The applicability of the ideas based on precedents rather than principles is doubtful, and the transferability of the terminology open to question. The appreciation of architectural qualities remains an essentially private visual language which is difficult to learn and probably irrelevant to apply. Perhaps the townscape school has achieved just such an application in transferring the discussion of visual elements to the level of townscape, but ambiguities and idiosyncratic interpretations abound. There is clearly something empty and superficial in the preoccupation with the visual, in the approach of the disembodied eye, in the objective approach of the professional evaluator or notationist. The visual quality paradigm tends towards the ultimate in terms of detached, external perception of the landscape.

Areas of common effort and consensus

These three paradigms encapsulate the major research traditions that impinge upon the experience of the urban landscape. They provide adequate vehicles for discussing and interrelating major bodies of work, but fail to provide profound insights or indications of where future research and synthesis is needed. They do suggest, however, areas where the cross-fertilisation of ideas from different fields is such that there appears to be substantial agreement. In a highly disparate field, such areas of consensus merit considerable emphasis.

The need is recognised increasingly for a wider concept of landscape, embodying physical settings, social behaviour and affixed meanings. Terms like 'townscape', 'place', '*genius loci*' and 'inscape' have been coined to express this fusion of traditional physical conceptions of landscape with the activities and meanings attached to them, and there is now conscious effort to synthesise place with community, townscape with street-life, and visual experience with conceptual and behavioural experience.

More significant still is the extent to which questions of meaning have become the major concern of landscape researchers, even for those who have sought objective evaluations or qualitative measures. From Jungian psychology has come an understanding of primordial symbolism and the collective unconscious; from anthropology information on the meanings attached to landscapes and settlements by primitive societies; and from ethology an understanding of the meanings of personal space and territory. From human geography, material has emerged on attitudes towards places and their effects on physical forms, and on how current economic and cultural forces express themselves in the landscape; from urban and architectural history insight into the intentions and aspirations of both those who actually shaped the land and built forms and those who consumed their product, and of how the *Zeitgeist* is expressed in tangible terms. From art and literature have been culled meanings of landscape and place to counterbalance and reinforce generalist interpretations, while introspective traditions have sought to provide the deep and presuppositionless meanings of original experience; and finally from the application of semiology to architecture has come initial understanding of signs and symbolism and their universality in landscape experience. All this interpretative material on the variety of meanings, and on the meaning of meaning, provides an essential counterweight to the two traditions of looking at the dynamics of the perception process and the visual landscape for their own sake. It provides that added dimension which reminds us of the complexity and irreducibility of the landscape experience.

It has been suggested that such interpretations and themes of meanings are parallel and complementary, but they also overlap significantly. In this way, they help validate one another and point to the utility of divergent approaches and research paradigms. The most substantive, if contentious, example is the degree to which neuropsychological theories of the brain can embrace and apparently reconcile *Gestalt* laws of forms, information theory, conceptions of perception, primordial and primitive symbolism, elementary semiology, and phenomenological and existential conceptions of space.[50] A second example might be found in interpretations of what modifications people make to their homes and gardens and of how they shape their personal landscapes. These interpretations began as a series of abstract aesthetic ideas, moved on to a discussion of fads and fashions in design, through discussions of cultural predispositions to expressions and presentations of self, beyond that to expressions of conscious motivation and response, until the ideas and actions are seen as the product of material forces within a particular historic context.[51] Depending on one's philosophical leanings, such interpretations may be seen as an interrelated hierarchy or as a set of partial explanations, but either way they contribute significantly to each other. This complementarity has led researchers to suggest that the strong traditions of eclecticism and cross-fertilisation within this field must be fostered and continued, that one must anticipate developments of interest and ultimate utility on many fronts, and that one must eschew the pursuance of one avenue to the exclusion of all others.

Yet the problem remains of how to integrate and synthesise these disparate strands. How can introspective findings be meshed with those of social survey? How can metaphysical statements from literary sources be balanced against contextual analysis of social consciousness? How can ideas derived from ethology be compared with those derived from social and environmental psychology? Any kind of rigorous synthesis must fall back upon the philosophical base of inquiry, in order to understand the starting point of different analyses, and their basic assumptions and methodological orientations. Perhaps certain philosophical traditions can be partially reconciled and the eclecticism noted above put on a firmer footing. Equally, diametrically opposed traditions can offer complementary insights by providing competing theories at two ends of the spectrum. The philosophy of aesthetics provides a rich vein of thought that attempts to answer many of the fundamental questions with which landscape writers have been wrestling for some time. It remains something of a mystery why so little work on landscape aesthetics makes significant reference to this literature.

Philosophy of aesthetics: alternative positions

In outlining the various historic and contemporary traditions in aesthetic theory,[52] it is noticeable that they all have some, albeit unconscious, representation in the literature on landscape. Even Classical and Renaissance theories of forms and Romantic extrapolations of human emotions upon landscape forms still have their adherents and influences.[53] Similarly, despite their rather rarified tone, associationist, idealist and metaphysical interpretations still exert influence, if only through the extent to which writings on landscape aesthetics are based upon selections from Croce, Santayana and Dewey.[54]

Of more potential relevance are semiological perspectives developed in linguistic analysis and now well established in architectural circles.[55] These perspectives have been responsible for profound, if conflicting, analytical approaches to the questions of the expressive meaning of the environment and its interpretation. Phenomenology, with its insistence upon the qualitative richness and irreducibility of experience, and an atheoretical openness of stimuli led to a significant reorientation of research in aesthetics and landscape interpretation. A focus on the intentionality expressed in landscape modification, and upon the experience of place as opposed to landscape, characterise this perspective[56] which is closely related to existentialism through the work of Husserl. Existentialist writers emphasise that it is man's existential life (life world) which is the 'given' of human experience that phenomenologists focus on, and thus questions of aesthetic meanings become questions of the meaning for man. In landscape aesthetics, the landscape is studied because it is the 'repository of human striving', providing indirect evidence of the human condition.[57]

Contemporary empiricist traditions within aesthetics are now well developed. Analytical aesthetics, re-examining the philosophical and linguistic bases of the subject, is only now having repercussions upon landscape aesthetics, through questioning its very theoretical and terminological base. Scientific empiricism has been concentrated primarily in psychological laboratory experimentation and is now well developed even with regard to landscape perception.[58] Empiricist approaches to art history have revived formalist conceptions and focused upon the morphology of art, the demarcation of styles and the canons of criticism.[59] While stylistic and morphological classification have been prominent in landscape aesthetics, they have been derided increasingly as avoiding the important questions of meaning, and debate on the canons of criticism has only taken place on a significant level in architectural history.[60]

While all these approaches within aesthetics are represented in the literature on landscape, generally they have been poorly developed. Far deeper levels of insight can be gained by applying conceptions and hypotheses derived from the analysis of aesthetic effects in art and literature. This in itself is a task worthy of further research, but there remains a more productive avenue in further developing the link between another philosophical position in aesthetics – dialectical materialism – and the study of landscape experience. The materialist perspective on aesthetics is not entirely new, for there are significant historical roots in the work of the Realists (Zola, Comte) and the Moralists (Tolstoy, Ruskin). When Marx and Engels formulated dialectical materialism, however, they did not apply it to aesthetic issues, except for a few literary works. They did not develop a well rounded aesthetic theory, and efforts to reconstruct such a theory have been fraught with problems, often degenerating into reductionist economism and naïve idealism.[61] Efforts to develop a Marxian aesthetic have been stunted by narrow interpretations of 'socialist realism' and preoccupation with the propagandist values of art and literature,[62] but now that wider and more thoroughgoing applications of materialist perspectives on art and literature are available, it is possible to see the potential in their applications to landscape aesthetics and environmental evaluation.

A materialist perspective: first steps in reorientation

In a sense, the fundamental reorientation induced by a materialist perspective has already been anticipated. In expressing dissatisfaction with narrow conceptions of the aesthetic, and searching through its socio-linguistic roots for the causes of the limited conception, this chapter has adopted a typical materialist approach embodied in existing aesthetic theory.[63] A materialist position also supports the wider conception of landscape and reaffirms the oft-demonstrated conclusion that in many instances landscape is merely a backdrop to experience. Functions, activities and people themselves often dominate place experience and environmental evaluation, especially within an urban context. These conventional separations of landscape from place, and of the aesthetic from other types of experience, are seen by materialists as part of the divided consciousness characteristic of modern Western society. As Williams[64] suggests:

we live in a world in which the dominant mode of production and social relationships teaches, impresses, offers to make normal and even right, modes of detached, separated, external perception and action: modes of using and consuming rather than accepting and enjoying people and things.

It is just such a mode of perception, the 'privileged indifference' of the critic, that must be overcome.

Several interrelated consequences emerge from this conclusion. First, landscape experience must be seen as part of everyday experience, varying according to the attitude, activity and knowledge of the observer. The abstraction of landscape experience to mere orientation or wayfinding, to pure visual contemplation, or to territorial demarcation reduces our understanding of the range and intentions of our responses so that we no longer understand even ourselves. Secondly, notions of the 'aesthetic moment' have to be related to the aforementioned range of our intentions and responses, not abstracted as something which is a specific property of an object, or which occurs only in response to particular qualities presented in a specific situation.[65] In other words, they have to be seen as clearly related to other forms of experience and more commonplace perceptions. It is necessary therefore to correct the traditional preoccupation with the transcendental and the topophiliac, and to throw more light on the less affecting and less self-conscious moments of landscape experience. A statement of praxis by Lowenthal and Prince[66] illustrates both current tendencies and the reorientation required:

> To understand environmental experience it is not enough to use scientific methods. Feelings and insights that transcend those of everyday life constantly infuse and enrich our awareness of the world around us. The roles of passion and mood are perhaps best explored through imaginative literature and the arts, which enlarge experience and epitomise styles of environmental organisation, preference and symbolism.

Here at once is the recognition of the task at hand. The role played by emotion, imagination, and knowledge in enriching environmental experience is acknowledged as is its immediate negation – the relegation of the everyday experience, the abstraction of the transcendental, the reliance on 'approved categories' of literature and art for higher level experience, which themselves become models for environmental perception.

This widespread mining of art and literature as a means of defining landscape experience tends to emphasise the tradition of the detached aesthetic. It plays down other components of direct experience such as tactile sensations of surface, motion and exertion, weather, social contact and 'unaesthetic' preoccupations of mood and thought. It ignores man's versatility and his capacity to react totally differently to phenomena presented under similar conditions, according to his intentions, preoccupations and current activities (his 'proteanism' to use the humanistic geographer's term for this capacity).[67] Perhaps recognition of less self-conscious and less abstracted ways of seeing and experiencing would invalidate, or make irrelevant, issues in environmental quality like 'poor aesthetic taste', 'visual blindness' or 'mass apathy' towards the visual environment.

A materialist perspective could also contribute towards resolving some of the moral questions posed by those who have attempted to reconcile aesthetic with socio-cultural judgements. Two quotations illustrate the issue. The first highlights Tuan's concern with the socially irresponsible judgements made by self-appointed landscape critics:

> Are we to evaluate only the surface appearance and forget the social and economic forces, often unjust, that have brought it into being . . . by 'visual blight', we may simply mean that we are judging the health of society on the ground of visual evidence in the landscape . . . the argument as to what makes for quality in an environment is pursued at the level of social and moral philosophy and of ecological principles, and not at the level of aesthetics.[68]

The second shows Jackson's similar reaction to the subjectivity of landscape criticism:

> much current environmental appreciation seems to lack the moral (or social) ingredient: the experience is vivid but shortlived. Far from revealing some 'universal' landscape quality which can serve as a model, most of the time it merely reveals the critic's own sensitivity and uniqueness. Instead of being an impressive *aggregate* of admirable qualities, beauty is defined as simply *one* aspect of an object which may in itself be objectionable.[69]

Both writers are seeking some means of replacing purely subjective, visual evaluations and appreciations with a more rigorous perspective rooted in social or moral philosophy. Tuan does not attempt to reconcile the aesthetic with the social and hence is forced to relapse into the 'higher order' metaphysics of beauty, denying the primacy of the social and moral. Jackson, though, seems to suggest just such a rigorous analysis of aesthetic situations and aesthetic moments as proposed by materialists. He notes the frequent narrowness of meaning and interpretation in 'the beautiful'; how it can contradict and conflict with other qualities and meanings, and how it tends to reveal more about the observer than the observed. Materialist perspectives would encourage the search for the 'universals' and 'aggregates' of beauty, but only via analysis of the specific situations and attitudes through which such feelings arise, paying particular attention to the extent to which such feelings conflict with and are contradicted by other sensations and emotions.

The very notion of evaluating only surface appearance and forgetting the social and economic forces and conditions – 'Pitying the plumage and forgetting the dying bird' as Jackson[70] expresses it – is anathema to materialists. It is their very quarrel with conventional notions of aesthetics, beauty, taste and visual quality. For them, qualitative judgements rest upon the ideological expressions and the reflected social realities displayed in the landscape, so that social and thereby moral perspectives are integral and explicit. Such is the power of the conventional aesthetic and its received categories of 'art', 'literature', 'architecture' and 'landscape quality', that in actual perceptions the sense of social realism often acts as a corrective device as the observer becomes conscious of the impulses which are defining awe, reverence or beauty. Near simultaneous feelings of anger, exploitation and expropriation can emerge in stark contradiction to 'aesthetic' emotions as an understanding of what such landscapes represent (directly *mean* in terms of the lives of the people who actually created, or who have to inhabit, these landscapes), and how and why they have been brought about (their means of production).[71] The 'beauties' of Las Vegas at night, the Manhattan skyline from afar, billboards or suburban ornamentation, English country houses and Georgian estates are all well documented as landscape 'treats' but can only be genuinely and purely enjoyed by the observer who is able to deny, or perversely acclaim, their social and moral content, their implications for human life, and abstract their purely visual appeal.[72]

Finally, the materialist position, and refinements to it developed in the context of cultural analysis, offer the means for more profound interpretations of landscape. The materialist perspective, as distinct from that of the idealist, allows us to ground the persistent landscape-expressed forms of consciousness (e.g. bourgeois individualism and suburbanism, landscape consumerism and tourism) in material processes. An emphasis upon cultural materialism permits a move away from the more traditional concerns with economic and narrowly-defined social productive forces as shaping agents. This allows explicit consideration of cultural forces (politics, law, education, literature, the arts and the mass media), but still treats these cultural forces as human-material.[73] It allows more emphasis to be placed upon human creativity and identity, and on the individual as against society than has traditionally been the case in more mechanistic materialist formations.

Using a materialist perspective as the means of synthesis for the literature on landscape aesthetics avoids a basic problem. Reference has already been made to hierarchical levels of analysis or partial explanation of landscape modification. Other methods and frameworks of interpretation consistently fail to relate their discovered formative processes or critical stances back to their fundamental roots. By limitation, abstraction or recourse to pure subjectivity or objectivity, they deny the complexity and contradictions inherent in the experience of the environment. Allusion has already been made to the possibilities of relating dialectically the specific findings of different research traditions, but the relationships and cross-fertilisations remain fragmented, contradictory and unresolved. When synthesis is attempted, there is relapse into even narrower traditions that are often a denial of both humanity and materiality. The newer forms of materialism[74] meet the need for a framework which recognises fundamental material forces of economy and society, and thereby culture, and which allows for human biology, individuality and subjectivity.

The potential of the materialist perspective may be illustrated by reference to the example of the varied interpretations to be found in Anglo-American writings on suburbia. Various writers have identified themes of personal expression, rural fundamentalism, consumerism and conspicuous consumption, community and social mobility, frontier ethic, privatism and now the all-encompassing 'suburbanism as a way of life'.[75] All these themes represent partial interpretations of the social significance or meaning of the suburban landscape and the forces behind its production. Such overlapping and contradictory conceptions can only be adequately synthesised through a thorough-going materialist framework, which dialectically relates the productive material forces and the social conditions of existence to the physical and social conventions of suburban life. Without such a perspective, analyses of suburban problems and, more dangerously, supposed diagnoses and cures remain at the symptomatic level – placelessness, rootlessness, isolation, loss of identity, lack of participation and urbanity – and this prevents real comprehension by critics, policy-makers and suburbanites themselves.

In fact, one can subject the literature on the suburban landscapes and its persistent themes to the same rigorous materialist analysis as can be applied to conventional notions of literature, analysing its *genres*, conventions, symbols, mediums, structures of feeling, residual and dominant hegemonic elements, abstracted traditions, formations, typifications, homologies and mediations all as expressions of particular class interests, particular social values and particular cultural practices.[76] These categories, however, do not merely provide a means of categorising the literature, they can be further applied to provide insights into the very production of that landscape and its actual significance. Thus in the suburban context the essential *typification* is the individual home on the individual lot, privately owned and occupied by the nuclear family, *homologous* to the atomisation of society and denial of community, collective guilt at this denial being expressed in certain *structures of feeling* such as rural fundamentalism. The essential *mediation* can be conceived as privatism, and the initial productive material force is the system of private ownership and capital accumulation fostered by the economic system, which requires wholesale suburbanisation as a means of increasing domestic consumption, overcoming crises of accumulation, and ameliorating the class struggle.[77]

This example is briefly and crudely drawn but it illustrates the complex hierarchy of levels of explanation mentioned earlier. The materialist perspective yields not only different levels of explanation, intersecting and reverberating with one another, but also a structure and logic for comprehending these reverberations. It provides not only a means for critical synthesis of partial conceptions, but also a level of interpretation directly related to the complexity and contradictions of human experience, in this context what Walker illuminatingly describes as the 'Faustian bargain' of the suburbs.[78]

Conclusion

In conclusion, four potential contributions of a materialist approach to landscape aesthetics in general may be suggested. First, the conceptions of both landscape and aesthetics are broadened. Aesthetic experience is seen as part of an indissoluble social experience, landscape is seen as an element of place inextricably bound up with social use and meanings. Together then, landscape aesthetics can represent a field of knowledge defined by its attempts to understand the nature of environmental experience without removing it from the context of everyday life.

Secondly, the materialist approach provides a framework for relating dialectically 'lower level' interpretations, whether these be biological, psychological or cultural bases of perception, or differing interpretations of the role of economic, social, technological or cultural factors in producing different landscape forms. The value of such a framework is that it provides a much more rigorous means of synthesis which is a major advance over the traditional eclecticism or the narrow paradigms thus far adopted.

Thirdly, the materialist position provides a qualitatively different position from which to criticise existing landscape literature and writing on landscape aesthetics. It provides a way of reformulating the real issues of concern (socio–moral), of demarcating the blind alleys, and establishing the route towards a fuller sense of (simultaneous) experience and interpretation. It has been a characteristic of materialist writing and aesthetics that its development and contribution has always been greatest in dialogue with and critique of opposing strains of thought — idealism, metaphysics, formalism, phenomenology.[79] For instance, the promising perspectives of hermeneutics, as a blend of phenomenology and materialism and 'political' semiology emerged from just such a dialogue.[80] There is a great deal to be gained from subjecting the literature on landscape aesthetics to the materialist critique and rewriting the history of landscape design and architecture to redress the balance of interpretation and to replace the exclusions. In this way, responses to different landscapes can be traced back to their social and historic (material) conditions. This complements the endeavour of tracing back given landscapes to their material and historical conditions of production. Through this double process, new understandings and evaluations have already emerged in other spheres of study.

Finally, to return to the very nub of the issue, it is crucial to understand the nature and meaning of what we conventionally call 'the aesthetic'. It is important to re-examine the conditions, attitudes and components of that particular group of experiences. The materialist approach insists that such experiences be examined within the context of all other experiences and meanings, not abstracted, studied and replicated self-consciously. In the landscape context, it is particularly important to look at less self-conscious traditions of landscape experience and enjoyment[81] and at the more incidental occasions when 'aesthetic' experience becomes almost submerged in a variety of utilitarian responses. As Williams notes in the context of literary analysis:

> Value cannot reside in the concentration or in the priority or in the elements which provoke these [experiences]. The argument of values is in the *variable* encounters of intention and response in *specific situations*. The key to any analysis, and from analysis back to theory, is then the recognition of precise situations in which what have been isolated and displaced as 'the aesthetic intention' and 'the aesthetic response' have occurred.[82]

The real understanding of environmental values and landscape aesthetics depends upon just such an analysis.

Notes

1 Cornish, V. 1928. Harmonies of scenery – an outline of aesthetic geography. *Geography* **14**, 275–83, 382–94.
Cornish, V. 1935. *Scenery and the Sense of Sight*. Cambridge: Cambridge University Press. For the writings of Cullen and the Townscape School, see notes 42 and 44. For the human geography writings, see notes 28, 29 and 30.
2 Appleton, J. H. 1975a. *The Experience of Landscape*. London: Wiley.
3 Williams, R. 1976. *Keywords: A Vocabulary of Culture and Society*, 27–8. London: Fontana/Croom Helm.
4 Houston, J. M. 1969. Editor's introduction. In *The World's Landscapes* series. London: Longman. Similar artistic origins can be traced in other European languages, e.g. Italian (*paeseggio*), Spanish (*paisagista*), French (*paysage*) and German (*Landschaft*).
See Gombrich, E. H. 1966. The renaissance theory of art and the rise of landscape. In *Norm and Form: Studies in The Art of the Renaissance*. London: Phaidon: and the discussion in Cosgrove, D. E. 1976. *The Urban Landscape of Vicenza*, 8–9. Unpublished D.Phil. Thesis. Department of Geography, University of Oxford.
5 Johnson, S. 1755. *A Dictionary of the English Language*. London: printed for J. Knapton and company.
For fuller discussion of the various meanings of 'Landscape' see Murray, J. A. and I. Murray 1903. *A New English Dictionary of Historical Principles*. Oxford: Clarendon Press.

6 Wood, L. J. 1970. Perception studies in geography. *Trans Inst. Br. Geog* **50**, 138. See also note 7.

7 Rapoport, A. 1977. *Human Aspects of Urban Form*. Oxford: Pergamon.

8 The distinction between vision (literal perception) and perception (schematic perception), the former being concerned with optical laws is made by Gibson, J. J. 1950. *The Perception of the Visual World*. Boston: Houghton Mifflin.

9 Rapoport, A. op. cit., 36–8.

10 Gibson, J. J. op. cit. and Gibson, J. J. 1966. *The Senses Considered as Perceptual Systems*. Boston: Houghton Mifflin.

11 Tunnard, C. and B. Pushkarev 1963. *Man Made America: Chaos or Control*, 173. New Haven, Conn.: Yale University Press.
Spreiregen, P. D. (ed.) 1967. *The Modern Metropolis: Selected Essays by Hans Blumenfeld*, 217–20. Montreal: Harvest House.
Lynch, K. 1971. *Site Planning*, 191–5. Cambridge, Mass.: MIT Press.
Hornbeck, P. 1968. *Highway Esthetics*. Cambridge, Mass.: School of Architecture, Harvard University.

12 Parr, A. E. 1965. City and psyche. *Yale Review* **55**(1), 71–85.
Wohlwill, J. F. 1976. Environmental aesthetics: the environment as a source of effect. In *Human Behaviour and Environment*, I. Altman and J. F. Wohlwill (eds), 37–86. New York: Plenum. This latter work is heavily based upon the work of Berlyne, D. E. 1972. *Aesthetics and Psychobiology*. New York: Appleton–Century. See also Berlyne, D. E. (ed.) 1974. *Studies in the New Experimental Aesthetics: Steps Toward an Objective Psychology of Aesthetic Appreciation*. New York: Halstead.

13 Rapoport, A. and R. E. Kantor 1967. Complexity and ambiguity in urban design. *J. Am. Inst. Planners* **33**, 210–21.
Lowenthal, D. 1972. *Environmental Assessment: A Comparative Analysis of Four Cities*. New York: American Geographical Society.
Lozano, E. E. 1974. Visual needs in the urban environment. *Town Plann. Rev.* **45**, 351–74.

14 Prak, N. L. 1977. *The Visual Perception of the Built Environment*. Delft: Delft University Press.
See also Crosby, T. 1973. *How To Play the Environment Game*, 74–81. London: Penguin.

15 Prak, N. L. 1977. op. cit., especially 46–57. Prak applies these conditional tendencies to explain levels of visual interest, differentiation of the streetscape, hard *vs.* soft/impenetrable *vs.* accessible facades and the like.

16 Strauss, A. L. 1961. *Images of the American City*. New York: Free Press.
Tuan, Y. F. 1974. *Topophilia: A Study of Environmental Perception, Attitudes and Values*, 192–224. Englewood Cliffs, NJ: Prentice-Hall.
Lynch, K. 1960. *The Image of the City*. Cambridge, Mass.: MIT Press.
Downs, R. M. and D. Stea 1977. *Maps in Minds: Reflections on Cognitive Mapping*, especially 209–39. New York: Harper and Row.

17 Prak, N. L. op. cit., 83–5.

18 Rozelle, R. M. and J. C. Baxter 1972. Meaning and value in conceptualising the city. *J. Am. Inst. Planners* **38**(2), 116–22.
Steinitz, C. 1968. Meaning and the congruence of urban form and activity. *J. Am. Inst. Planners* **34**(4), 233–48. See also Prak, W. L. 1977. op. cit., 83–5.

19 Rapoport, A. op. cit., 65–80.

20 Prak, N. L. op. cit., 83–90. See also note 19.

21 Smith, P. F. 1974. *The Dynamics of Urbanism*. London: Hutchinson.
Smith, P. F. 1977. *The Syntax of Cities*. London: Hutchinson.
Smith, P. F. 1979. *Architecture and the Human Dimension*. London: George Godwin.

22 For a review of these findings see Rapoport, A. op. cit., 277–89.

23 Appleton, J. H. op. cit.
24 See, for example, Newman, O. 1973. *Defensible Space*. London: Architectural Press.
   Becker, F. D. 1977. *Housing Messages*. Stroudsburg, Pa: Dowden, Hutchinson & Ross.
   Lancaster, O. 1958. *A Cartoon History of Architecture*. Boston: Houghton Mifflin.
   Venturi, R. and C. Rauch 1976. Signs of life. *Arch. Des*. **46,** 496–8.
   Werthman, C. S. 1969. *The Social Meaning of the Physical Environment*. Unpublished Ph.D. thesis, Department of Sociology, Berkeley, University of California.
25 See the comments of Helen Ross, quoted in Canter, D. and P. Stringer (eds) 1975. *Environmental Interaction*, 5. London: Surrey University Press.
26 Tuan, Y. F. 1971. Geography, phenomenology, and the study of human nature. *Can. Geogr*. **15,** 181–92.
27 Sauer, C. O. 1925. The morphology of landscape. Reprinted in J. Leighly (ed.) 1963. *Land and Life: A Selection from the Writings of Carl Ortwin Sauer*, 315–50. Berkeley: University of California Press.
28 Tuan, Y. F. 1974. *Topophilia: A Study of Environmental Perception, Attitudes and Values*. Englewood Cliffs, NJ: Prentice-Hall.
   Tuan, Y. F. 1977. *Space and Place: The Perspective of Experience*. London: Edward Arnold.
29 Lowenthal, D. and H. C. Prince 1964. The English landscape, part 1. *Geog. Rev*. **54,** 309–46.
   Lowenthal, D. and H. C. Prince 1965. The English landscape, part 2. *Geog. Rev*. **55,** 186–222.
   Lowenthal, D. 1968. The American scene. *Geog. Rev*. **58,** 61–8.
30 See the comments of J. B. Jackson and Y. F. Tuan, in Lewis, P. H., D. Lowenthal and Y. F. Tuan (eds) 1973. *Visual Blight in America*, 23–7, 47–8. College Resource Paper 23. Washington: Association of American Geographers.
   See also Zube, E. H. (ed.) 1970. *Landscapes: Selected Writings of J. B. Jackson*. Amherst: University of Massachusetts Press.
31 Relph, E. C. 1976. *Place and Placelessness*, 46–7. London: Pion. Relph derives his definition of place from the writings of Camus.
32 Martin, G. H. 1968. The town as palimpsest. In *The Study of Urban History*, H. J. Dyos (ed.), 155–70. London: Edward Arnold.
   Hoskins, W. G. 1955. *The Making of the English Landscape*. London: Hodder & Stoughton.
33 See the comments of R. Lubove 1977. Review of books. *Urban History Yearbook 1976*, 95. Leicester: Leicester University Press.
   Also Editorial. In *Urban History Yearbook 1977*, 2–23. Leicester: Leicester University Press.
34 Sewell, J. 1974. *A Sense of Time and Place*, 4–5. Toronto: City Pamphlets.
35 Allsopp, B. 1970. *The Study of Architectural History*. London: Studio Vista.
   Open University 1975. *History of Architecture and Design 1890–1939*. Milton Keynes: Open University Press.
36 See for example Gloag, H. 1962. *Victorian Taste*. Newton Abbot: David & Charles.
   Briggs, A. 1977. A cavalcade of tastes. *Arch. Rev*. **162,** and successive issues.
   Jordan, R. F. 1966. *Victorian Architecture*. London: Pelican.
37 MacCormack, R. 1978. Housing and the dilemma of style. *Arch. Rev*. **163,** 203–6.
   See also Benton, T. C. 1975. Architecture and building. In Open University op. cit., 32.
38 Gombrich, E. H. 1975. *Art History and the Social Sciences,* 57. Oxford: Clarendon Press.
39 The problems which bedevil a coherent approach to architectural semiology are

concisely summarised in Agrest, D. and M. Gandelsonas 1977. Semiotics and the limits of architecture. In *A Perfusion of Signs*, T. A. Seboek (ed.). Bloomington: Indiana University Press. This paper also considers the application of semiotic perspectives to the built environment at large.

See also Broadbent, G. 1976. A plain man's guide to the theory of signs in architecture. *Arch. Des.* **47**, 474–82.

40 See for example Scruton, R. 1979. *The Aesthetics of Architecture*, 158–78. London: Methuen.

41 The different approaches towards architectural semiology are well represented in the contributions included in Broadbent, G., R. Bunt and C. Jencks (eds) 1979. *Signs, Symbols and Architecture*. Chichester: John Wiley. See especially the papers by U. Eco and G. Broadbent. Comparison should be made between this volume and its predecessor (Jencks, C. and G. Baird (eds) 1969. *Meaning in Architecture*. London: Barrie and Rockcliffe) to assess the direction and progress of research in architectural semiology.

42 Cullen, G. 1971. *The Concise Townscape*. London: Architectural Press.

43 Taylor, N. 1973. *The Village in the City*, 17. London: Temple Smith. Taylor is specifically criticising works like Wolfe, I. de 1963. *The Italian Townscape*. London: Architectural Press.

44 Relph, E. C. op. cit.
Whistler, W. M. and D. Reed 1977. *Townscape as a Philosophy of Urban Design*. Exchange Bibliography 1342, Monticello, Ill.: Council of Planning Librarians.

45 Maxwell, R. 1976. An eye for an I: the failure of the townscape tradition. *Arch. Des.* **46**, 534–6.

46 Thiel, P. 1961. A sequence-experience notation for architectural and urban spaces. *Town Planning Rev.* **32**, 33–52.
Appleyard, D., K. Lynch and J. R. Myer 1964. *The View from the Road*. Cambridge: MIT Press.
Cullen, G. 1967. *Notation 1, 2 and 3*. Banbury: Alcan Industries.

47 Penning-Rowsell, E. C. 1974. Landscape evaluation for development plans. *The Planner* **60**, 930–4.
Robinson, D. G., J. F. Wager, I. C. Laurie and A. L. Traill 1976. *Landscape Evaluation*. Manchester: University of Manchester, Centre for Urban and Regional Research.

48 Appleton, J. H. 1975b. Landscape evaluation: the theoretical vacuum. *Trans Inst. Br. Geogs* **66**, 120–3.

49 See for example Reekie, R. F. 1972. *Design in the Built Environment*. New York: Crane Russak.
Logie, G. 1954. *The Urban Scene*. London: Faber & Faber.

50 Smith, P. F. 1977. op. cit.

51 See respectively Watts, M. T. 1957. *Reading the Landscape*, 197–220. New York: Macmillan.
Jackson, J. B. 1969. Ghosts at the Door. In *The Subversive Science: Essays Towards an Ecology of Man*, P. Shepard and D. McKinley (eds), 158–68. Boston: Houghton Mifflin.
Werthman, D. op. cit.
Becker, F. D. op. cit.
Veblen, T. 1899. *The Theory of the Leisure Class*. New York: Macmillan.

52 The identification of various traditions in aesthetics is drawn from Beardsley, M. C. 1966. *Aesthetics: From Classical Greece to the Present*. Alabama: Alabama University Press.

53 For examples see Hillier, B., A. Leaman, P. Stansell and M. Bedford 1976. Space

syntax. *Environment and Planning B* **3**(2), 147–85; and Shoard, M. The Lure of the Moors in this volume.
54 Such interpretations tend to be preoccupied with intuition, emotion, imagination, intellectual sympathies and the 'hushed reverberations of the mind'. See for example Appleton, J. H. 1975b op. cit, Lowenthal, D. and H. C. Prince 1977 infra cit.
55 See note 41.
56 See notes 28 and 31.
57 See note 26: and
Norberg-Schultz, C. 1971. *Existence, Space and Architecture*. London: Studio Vista.
58 See note 12.
59 Beardsley, M. C. 1966. op. cit., 376–88.
60 Allsopp, B. op. cit. especially 98–105.
See also Attoe, W. 1977. *Architecture and Critical Imagination,* especially 11–108. New York: Wiley.
Gauldie, S. 1969. *The Appreciation of the Arts: 1. Architecture*. London: Oxford University Press.
Scruton, R. op. cit.
61 Baxandall, L. and S. Morawski 1974. *Karl Marx, Frederick Engels on Literature and Art,* 3–47. New York: International Graphic.
62 Beardsley, M. C. op. cit., 355–63.
Laing, D. 1978. *Marxist Theory of Art*. London: Harvester.
63 Laing, D. op. cit., 39.
64 Williams, R. 1975. *The Country and the City,* 358. London: Paladin.
65 Williams, R. 1977. *Marxism and Literature,* 151–7. Oxford: Oxford University Press.
Mukarovsky, J. 1970. *Aesthetic Function, Norm and Value as Social Facts,* especially 1–23. Ann Arbor: University of Michigan, Department of Slavic Languages and Literature.
66 Lowenthal, D. and H. C. Prince 1977. Transcendental Experience. In *Experiencing the Environment,* S. Wapner, S. B. Cohen and B. Kaplan (eds), 117. New York: Plenum.
67 Relph, E. C. op. cit., 122–35.
68 Tuan, Y. F. 1973. Visual blight: exercises in interpretation. In P. F. Lewis, D. Lowenthal and Y. F. Tuan op. cit., 26.
69 Jackson, J. B. 1973. Commentary: visual blight – civil neglect. In P. F. Lewis, D. Lowenthal and Y. F. Tuan op. cit., 47.
70 Jackson, J. B. 1967. To pity the plumage and forget the dying bird. *Landscape* **17**(1), 1–4.
71 New Left Review 1979. *Raymond Williams: Politics and Letters,* 345–9. London: New Left Books.
Williams, R. 1975. op. cit., especially 131–4 and 179–80.
72 See for example Venturi R. and D. S. Brown 1968. *Learning from Las Vegas*. Cambridge, Mass.: MIT Press.
Goodman, R. 1972. *After the Planners,* especially 164–74. London: Pelican.
Kurtz, S. 1973. *Wasteland, Building the American Dream*. New York: Praeger.
See also note 70.
73 Williams, R. 1977. op. cit., 90–4.
74 Williams, R. 1973. Base and superstructure in Marxist cultural theory. *New Left Rev*. **82,** 3–16.
Williams, R. 1978. Problems of materialism. *New Left Rev*. **109,** 3–17.
Timpanero, S. 1975. *On Materialism*. London: New Left Books.

75 See for example Banham, R. 1971. *Los Angeles: The Architecture of Four Ecologies*. London: Penguin.
   Taylor, N. 1973. op. cit.
   Berry, B. J. L. 1976. *Urbanization and Counter Urbanization*, 17–30. Beverley Hills: Sage.
   Hall, P. G. 1969. The urban culture and the suburban culture. In *Man in the City of the Future*, R. Eells and C. Walton (eds). London: Collier-Macmillan.
   For specific elements in these interpretations see also notes 51, 72, 77.
76 Williams, R. 1977. op. cit., 90–191.
77 Pawley, M. 1973. *The Private Future*. London: Thames and Hudson. Harvey, D. 1978. Labor, capital and class struggle around the built environment in advanced capitalist societies. In *Urbanisation and Conflict in Market Societies*, K. Cox (ed.), 9–37. London: Methuen.
   Walker, R. A. 1977. *The Suburban Solution: Urban Geography and Urban Reform in the Capitalist Development of the United States*. Unpublished Ph.D. thesis, Department of Geography, Baltimore: Johns Hopkins University.
78 Walker, R. A. op. cit., 608–15.
79 Beardsley, M. C. op. cit. 355–63.
80 See for example Wolff, J. 1975. *Hermeneutic Philosophy and the Sociology of Art*. London: Routledge & Kegan Paul.
   Barthes, R. 1976. *Mythologies*. London: Paladin.
81 Examples of such traditions may be found in more physically-active and more socially linked forms of recreation and leisure where landscape 'experience' is less self-consciously sought. An illuminating example is Hugill's description of the seafront of Southend, England. In Hugill, P. J. 1975. Social conduct on the Golden Mile. *Ann. Assoc. Am. Geogs* **65**, 214–28.
82 Williams, R. 1977. op. cit., 157.

7 Humphry Repton and the morality of landscape

STEPHEN DANIELS

Introduction

The landscape parks of Georgian England impress us with their composure – a quality that distinguishes many aspects of polite Georgian culture – but landscaping was a more contentious practice than these impressions may suggest. In a culture in which appearances codified social attitudes and relationships to a remarkable degree,[1] there would be as much argument about the design of a park as there was about the details of personal dress, bearing and expression. Debates on the style of landscape parks were not just about their formal aesthetic qualities. In a period when the gentry were at the height of their power, the landed estate was seen as a metonym of English society and changes to its landscape raised social issues, especially the moral questions of rights and duties.[2] This essay investigates the varying and often conflicting ways that Georgian writers on landscape gardening formulated the relationship between aesthetic values and moral values. It focuses upon the work of Humphry Repton, a practising landscape gardener who also wrote at length about this relationship, and in particular, analyses his commission at Sheringham in Norfolk, where, at a time of acute class tension, he designed a landscape to improve both the appearance and social relations of an estate.

Landscape and morality

Throughout the Georgian period, Tory moralists denounced a type of landowner – grasping, speculative, ambitious and profligate – that they associated with the Whigs. They reaffirmed the conservative image of a settled, cultivated countryside in the care of an old landed family in response to the growing number of large, emparked estates owned, but only occasionally occupied, by the new Whig gentry.[3] 'Estates are landscapes', complained William Cowper, 'gaz'd upon a while, then advertize'd and auctioneered away'.[4] The most celebrated indictment of Whiggish landscaping is Oliver Goldsmith's *The Deserted Village,* first published in 1770.[5] The poem describes the destruction of a community of peasant proprietors by an incoming nabob. The ruins of the village, including the parsonage where the 'long remembered beggar' and 'broken soldier' were welcome, lie beneath his landscape park.

> The man of wealth and pride
> Takes up a space that many poor supply'd,
> Space for his lake, his park's extended bounds,
> Space for his horses, equipage and hounds . . .
> His seat where solitary sports are seen,
> Indignant spurns the village from the green.

Goldsmith connects the social ruin of the country with its growing radiance, a process which culminates in dazzling displays of power.

> But verging to decline its splendours rise,
> Its vistas strike, its palaces surprise . . .
> Ye friends to truth, ye statesmen who survey
> The rich man's joys increase, the poor's decay,
> 'Tis yours to judge how wide the limits stand
> Between a splendid and happy land.

Landed property, which underwrote the power of the gentry, was protected by an increasingly harsh penal code. A Whig Act of Parliament made it a capital crime to destroy the trees in a gentleman's park or poach his game.[6] Goldsmith observed how 'possessions are paled up with new edicts every day and hung round with gibbets to scare invaders'.[7] The love of property was manifest in landscape tastes. In extreme versions of the picturesque, where land was reduced to scenery and manipulated as such, self-conscious observation was inseparable from self-conscious ownership.[8]

Not all conservative criticism of Whig landscapes implied rejecting the practice of landscaping. Alexander Pope argued for landscape gardens that expressed a morality of 'sense', in which the bourgeois virtue of utility was combined with and tempered by the patrician virtue of charity.[9] Following William Gilpin, many picturesque theorists considered the moral faculty to be separate from and often contradicted by the pictorial imagination,[10] but the Herefordshire squire Uvedale Price proposed a picturesque aesthetic that was informed by benevolence. Like many conservatives, he abhorred the style of Lancelot 'Capability' Brown and his imitators which isolated country houses in depopulated parks. He wished to restore a sociable landscape with distinctions but not divisions, based on an aesthetic and social principle he termed 'connection'.[11] In 1795, he wrote:

> Although the separation of the different ranks and their gradations, like those of visible objects, is known and ascertained, yet from the beneficial mixture and frequent intercommunication of high and low, that separation is happily disguised, and does not sensibly operate on

the general mind. But should any of these links be broken; should any distinct undisguised line of separation be made, such as between noble and roturier, the whole strength of that firm chain (and firm may it stand) would at once be broken.[12]

Price described landscape gardening as an 'art of peace' at a time when many of the gentry were experiencing acute feelings of disquiet. They were anxious at industrial developments, the erratic behaviour of the economy, soaring poor rates, the threat of rebellion at home and invasion from abroad.[13] Profoundly disturbed by the French Revolution, Price gave a warning to those whose vast possessions isolated them from the commonalty. He advised estate owners to pay:

> increased attention to that most necessary and useful body of men – the labourers: for without their genuine attachment, however firmly we may be united to each other, our union would be far from complete . . . he who can scarcely buy bread will hardly buy arms unless driven to despair by ill treatment.[14]

Price did not question the sanctity of property. His complaint, and that of many who still pictured their social world as a hierarchy of ranks, was that property had been divorced from duty. They appealed for the restoration of patrician responsibilities, rarely for plebeian rights.

Humphry Repton

From 1790 until his death in 1818, Humphry Repton was the leading landscape gardener in England.[15] This period was one of radical change in many aspects of English life and it is therefore not surprising that Repton's philosophy of landscape, which engaged such issues as enclosure, industrialisation, poverty and class conflict, was changeful, complex and ambiguous. Repton's own ideas are not always revealed in his commissions. In the Red Books prepared for individual commissions he had to accommodate the tastes of his clients; moreover, the landscape that materialised sometimes bore little resemblance to the one Repton designed. This galled Repton and in his published works he seized the opportunity to express his own ideas and criticise those of some of his clients.

Repton first practised his art on his own small estate when he lived at Sustead in Norfolk between 1778 and 1786. Here he also gained some personal regard for rural responsibilities. 'It is so small a parish', he wrote to Edward Chamberlayne, 'that I am obliged to enact the various parts of churchwarden, overseer, surveyor of the highways and esquire of the parish . . . I am impatient to show you the alteration to my house and grounds. The wet hazy meadows, which were deemed incorrigible have been drained and transformed to flowery meads.'[16] During his residence at Sustead, Repton sketched and painted the surrounding countryside, especially the local parks and country houses, and made a study of botany. He enjoyed but could not afford this gentlemanly life and in 1786 moved to a small cottage at Hare Street in Essex. In 1788, aged 36, he decided to make his leisure profitable and become a professional landscape gardener.

Repton read the literature on landscape gardening and singled out William Gilpin, William Mason, Thomas Whateley and René Louis Gerardin as his formative influences.[17] Of these writers, only Gilpin disassociated landscape aesthetics and social morality. Mason argued for paternal gardens that combined beautiful and useful scenes and revealed villages.[18] Whateley valued cultivated and populous landscapes that both expressed and stimulated benevolent feelings.[19] Gerardin believed that the 'virtuous citizen' who retired to the country to enjoy nature would 'soon feel that the sufferings of humanity make the most painful of all spectacles; if he begins by the admiration of picturesque landscapes which please the sight, he will soon seek to produce the moral landscapes which delight the mind.'[20] Repton visited the parks designed by Capability Brown and studied plans Brown made for his commissions.[21] Repton admired Brown and regarded himself as his successor, but there are pronounced discontinuities between their work. The scale of Repton's work and the wealth of his clients were modest by comparison. He did not agree with Brown's practice of isolating houses in acres of smooth greensward. Repton's ideal landowner was not a Whig grandee but an established Tory squire: this despite, or perhaps because, few of his clients actually fitted this description. The style of landscaping which most impressed and influenced Repton was that practised by his friend Norton Nicholls. On a visit in July 1788 to the park Nicholls designed for Sir William Jermingham at Costessy in Norwich, Repton wrote some appreciative verse in the guest book. He expressed his liking for the pictorial qualities of the park but particularly admired the way the landscape features symbolised Jermingham's virtues of sociability, charity and hospitality.[22]

In 1794 Repton published extracts from his first commissions together with commentary under the title *Sketches and Hints on Landscape Gardening*. He listed 16 sources of pleasure in landscape gardens, of which 'picturesqueness' was but one and not even the most important.[23] After four years' practical experience as a landscape gardener, Repton was disenchanted with the picturesque and now recognised much less affinity between landscape gardening and landscape painting.[24] He saw himself as the practitioner of an essentially useful art. The places he improved were primarily to be lived in, not looked at. He regarded the prevailing pictorial conception of landscape as too narrow and appealed to an earlier, more inclusive definition as 'a region or prospect of a country'. Unlike many picturesque theorists, he did not consider landscape taste as purely sensual. He endorsed Burke's argument that taste was informed by reason as well as imagination and concluded that good landscape taste was ultimately a matter of understanding.[25] He became impatient with attempts to disguise the use of the landscape to achieve pictorial effects, and admired bridges, textile mills, prisons and asylums which looked authentic.[26]

Repton never actually abandoned picturesque conventions, and the imperatives of beauty and use often co-exist in unresolved tension in his work. In his earlier works Repton argued that the sense of freedom a gentleman enjoyed when he looked over his park was threatened by the obtrusion of vulgar land uses:

> The pleasure of appropriation is gratified in viewing a landscape which cannot be injured by the malice or bad taste of a neighbouring intruder: thus an ugly barn, or ploughed field, or any obtrusive object which disgraces the scenery of a park, looks as if it belonged to another, and therefore robs the mind of the pleasure derived from appropriation, or unity and continuity of unmixed property.[27]

Picturesqueness and property are here inseparable and it is not incidental that Repton uses a metaphor of criminal theft to describe how the enjoyment of scenery is diminished. In his second treatise *Observations on the Theory and Practice of Landscape Gardening,* published in 1803, Repton expressed his admiration for useful landscapes which proclaimed the virtue of industrious-ness and generated revenue, but he did not think this justified admitting them into a gentleman's purview. He did not doubt that farmers preferred the prospect of a full crop of wheat or of livestock fattening within small enclosures to the more informal and more wasteful scenery of a park, but this outlook did not become the gentleman who owned the farmland. Quoting Pope, Repton thought such landowners should see those places 'where cheerful tenants bless their yearly toil but only occasionally and on condition they go outside their parks'. He summarised: 'It is the union, not the existence, of beauty and profit, of laborious exertion and pleasurable recreation, against which I would interpose the influence of my art'.[28]

Repton was sensible to the social implications of landscaping. In his *Enquiry into the Changes of Taste in Landscape Gardening,* published in 1806, he contrasted the ways in which the environs of 'ancient' and 'modern' country houses expressed the status of their owners, thus:

> When we formerly approached the mansion through a village of its poor dependants we were not offended at their proximity because the massy gates and numerous courts sufficiently marked the distance between the palace and the cottage . . . the village, the almshouse, the parish school and churchyard, were not attempted to be concealed by the walls and palisades that divided them from the embellished pleasure ground.

Repton noted how the modern taste for landscape was incompatible with the existence of these walls, which produced a gloomy and confined outlook, and in consequence they were frequently demolished. This had the effect of revealing villages to be uncomfortably close. Repton only condoned the practice of removing or concealing them as a last resort.[29] It was not necessary, he thought, to depopulate the land to enhance the importance of a country house. He advised his client Edward Foley to provide comfortable cottages for the labourers on his estate but at a respectful distance from his house:

> It is no more necessary that these habitations should be seen immediately near the palace than their inhabitants should dine at the same table; but if their humble dwellings can be made a subordinate part of the general scenery they will, so far from disgracing it, add to the dignity which wealth can derive from the exercise of benevolence.[30]

Repton's advice on the placement of labourers' cottages was consistent with his advice on the treatment of their inhabitants. While he welcomed providing them with 'wholesome food, warm clothing and good instruction' he thought it folly to teach them to read and write: 'I contend that some degree of ignorance is necessary to keep them subordinate'.[31]

Repton's last published work, *Fragments on the Theory and Practice of Landscape Gardening* (1816), is as much a moral tract as an aesthetic treatise. Repton recognises but now regrets that 'appropriation, the *exclusive right* of enjoyment, is a propensity that is part of human nature'. Throughout his career he had met few clients who preferred the sight of men to that of cattle, lawns and woods. The greed of the newly rich had transformed mature and benevolent landscapes into raw and brutal ones. In the *Fragment* ironically entitled *Improvements* he describes and illustrates the changes he witnessed when an old, established landlord sold his estate (Fig. 7.1). The common was enclosed, rents doubled, oak trees were uprooted and replaced by conifers, a bench for villagers to rest and admire the scenery was removed, a public footpath through the park was stopped up, the stile was removed and replaced with a warning about man traps and spring guns, the old mossy paling was knocked down and replaced by a new high fence 'not to confine the deer,' observed Repton 'but to exclude mankind'. The landscape was, so to speak, demoralised. In the *Fragments* Repton advocated improvements which would maintain or restore the qualities which once characterised this estate.[32] The commission at Sheringham, which he quotes at length, was his last major work and his most sustained attempt to incorporate moral notions in landscape design.

IMPROVEMENTS

Figure 7.1   Repton's ironical view of landscape improvement.

Sheringham

Repton considered his commission at Sheringham in 1812 the culmination of
his art. He found it particularly agreeable because the site was the most
promising he had seen and because, unusually in his experience, his client,
Abbot Upcher, shared his tastes and opinions. Upcher purchased the manor
of Sheringham from Cook Flower in July 1811. He first met Repton a few
weeks later in the offices of Repton's son William, the solicitor for the sale, at
Aylsham. For five days in June the following year Upcher and Repton
wandered together over the estate to assess its potential. The Red Book,
prepared in July 1812, records the substance of their conversations.[33]

The parish of Sheringham stretched down from the Cromer Ridge to the
North Sea. It contained a remarkable variety of topography and included
both a farming village and a fishing village. Repton was no stranger to its
attractions. Sustead was only four miles away and when living there he
contributed an illustration and description of Sheringham to a History of
Norfolk published in 1781. In it, he had singled out the estate that he was to
landscape 31 years later:

> Upper Sheringham is beautifully adorned by the extensive woods of
> Mr. Cook Flower, the summits of the hills are planted, whilst their
> bottoms and the rich valleys that divide them are variegated with
> unenclosed arable land which, though the soil is light, produces
> excellent barley, wheat and turnips.

Repton admired the coastal scenery also – the expansive views of the open
sea and the shoreline scenes of fishermen hauling up boats, landing their
catch, drying nets and repairing tackle.[34] In the late 18th and early 19th
centuries the area attracted an increasing number of tourists and artists. When
he passed through Sheringham in September 1798, Samuel Pratt enthused
over the scenery, repeating verbatim some of Repton's description, but
remarked on the contrast between the richness of the cultivated fields and the
poverty of the villagers in their 'miserable looking huts'.[35] Edmund Bartell, a
local surgeon and amateur water colourist, made the same observation in a
popular guidebook first published in 1800.[36] Bartell advocated landscapes
which proclaimed the benevolence of the rich and the contentment of the
poor and himself designed some model cottages which were both comfortable and picturesque. He saw a darker side to Norfolk's agricultural
prosperity. Like many observers he feared that the balance of its rural society
was upset by the large farmers with their growing wealth, social pretensions
and mean spirited attitude to the poor. He repeated the appeal to landowners
to intervene on behalf of an impoverished and disaffected labour force and
thereby restore social harmony.[37] This sense of alarm coloured Repton's

perception of Sheringham when he visited again in 1812, and the appeal for paternalism informed the new landscape that he envisaged.

The social climate of Norfolk became appreciably harsher during the Napoleonic Wars. The expansion of cereal cultivation increased the affluence of farmers and intensified the transformation of farm servants into casual labourers. Wages failed to keep pace with the rising price of provisions. The year of Repton's commission at Sheringham (1812) was one of peak prosperity for Norfolk farmers, taking full advantage of the high price of wheat, and one of chronic distress for farm labourers scarcely able to buy bread. Prosperous cereal farming was both a consequence and a cause of the increased pace of enclosure, mainly of commons and wastes. Agricultural improvers regarded commons as places where crime and idleness flourished and recommended enclosure as an instrument of social discipline. Expenditure on poor relief soared, reaching peaks in famine years. The Poor Laws were the subject of bitter controversy. The new incorporated work-houses usually provided more material comfort than old parish poorhouses, but in the eyes of the poor they symbolised the institution of a punitive regime and in the eyes of paternalists the failure of local, personal class relations. Many parishes employed variants of the Speenhamland System of outrelief in which subsistence wage levels were brought up to an agreed level by a supplement from the parish. Farmers took advantage of the system by paying low wages. For labourers of Speenhamland parishes it became immaterial whether they worked or not. They were at least guaranteed an income and had a vested interest in maintaining things as they were, which may explain why, in a period of intense suffering, many labourers were not openly hostile towards the authorities. Nevertheless, food riots were frequent and some had a political edge that alarmed the authorities. Poaching, often in gangs, became endemic in many parts of Norfolk. In a period when the gentry were improving methods of rearing, preserving and shooting game, and hungry labourers saw the commons appropriated, it is not surprising that poaching hardened class antagonism.[38]

What were the significant developments in Sheringham? The Enclosure Bill was passed in 1809 and the Award made in 1811, shortly before Abbot Upcher purchased the estate. The Land Tax Assessments do not suggest there was an appreciable redistribution of property and as far as the poor were concerned the award was more lenient than many. Two areas of common were taken in and two allotments created where they were allowed to cut furze and dig turfs.[39] It is questionable whether they thought this adequate compensation. There is no record of popular protest, but a labourer con-victed of poaching at nearby Holt may well have voiced a general complaint when he protested that, before the enclosure of common land there in 1812, he could hunt rabbits at will.[40] The expenditure on poor relief in the parish rose and fell in accordance with that for the county. In 1805 Sheringham combined with five neighbouring parishes in a Gilbert Union and a new

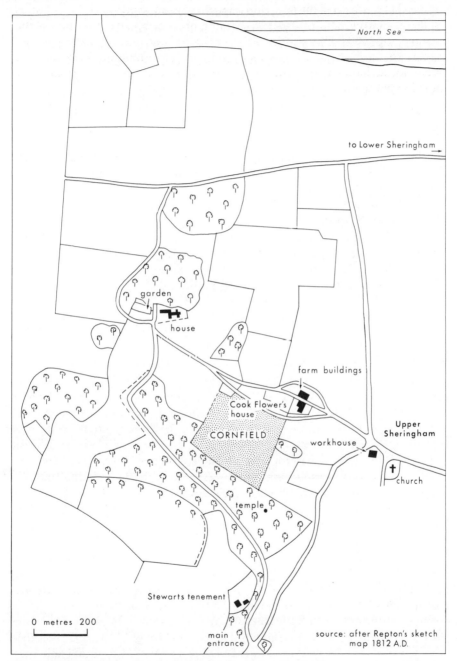

Figure 7.2 Humphry Repton's plan for the park at Sheringham.

workhouse (Fig. 7.2) serving the union was built in Upper Sheringham on Cook Flower's estate.[41] It was essentially an asylum for the old, the infirm, their dependent children and orphans; built to accommodate 150, it was probably a material improvement on the parish poorhouse built in the 1790s. We might infer from what we know that social conditions in Sheringham were not as harsh as in some parts of Norfolk; yet this is not an adequate position from which to evaluate the observations and proposals Repton made on his visit in 1812. The Red Book is not a direct commentary on events in the parish. Repton saw Sheringham as an instance of conditions prevailing generally in Norfolk and over much of rural England; moreover his way of seeing was decisively influenced by his reading of Pope, Goldsmith and George Crabbe,[42] and conversations with Abbot Upcher, his client.

Abbot Upcher and his wife Charlotte were fitting stewards for the landscape Repton envisaged. They combined an appreciation of the scenic beauties of Norfolk with an earnest benevolence. When they lived at Thompson, near Swaffham, Upcher supervised his home farm, visited the poor and dispensed alms. He and his wife read the Bible to their servants and invited the destitute to dine in their kitchen. 'I resolve never to turn a beggar from the door without giving him some relief', Upcher wrote in his diary in April 1812. When the Upchers purchased the Sheringham estate three months later they planned to thank God by 'doing all the good which lies in our power to the poor and needy of Sheringham and by setting a virtuous example to all around us'.[43]

The estate cost Upcher 50000 guineas. Repton intended to improve its value but not just to satisfy 'those who know no standard of value but Gold and its flimsy representatives'.[44] Repton questioned the abstraction of aesthetic as well as economic values from the land: 'mankind are apt to judge by the *eye* rather than the *understanding*', he commented, 'and more commonly select objects for their beauty than for their use or intrinsic worth'. His design was for a responsible, resident landowner who was closely involved in the running of his estate; yet the look of the land was important because it would reflect these virtues and provide a satisfying and encouraging image for Upcher and his guests to contemplate. In England generally during the Napoleonic Wars, and particularly in Norfolk, there was a growing taste for landscape which included evidence of work, organisation and management. Abbot Upcher had an informed interest in agricultural improvements and admired the cultivated appearance of Thomas Coke's estate at Holkham; 'the fields are like gardens' he wrote.[45] Repton proposed admitting a cornfield into the principal view from the new house (Fig. 7.3) and ridiculed those landowners with fastidiously picturesque tastes who recoiled from such a sight.[46] The cornfield would, he foresaw, 'lend some variety to the colouring of the picture and at seed time and at harvest time may be enlivened by men as well as beasts – if I may be permitted to indulge my favourite propensity for *humanising* as well as *animating* beautiful

Figure 7.3 View from the new house at Sheringham.

scenery'. Cook Flower's residence, which Upcher disdained as 'only a better kind of farmhouse'[47] was close to the cornfield but badly positioned when the estate was considered as scenery. Repton implied that Flower was too involved in the running of his farm. Although Repton found the combination of woodland, pasture and arable fields attractive, he remarked that land use was dictated by purely utilitarian considerations. He advised converting arable land near the site of the new house to pasture to create a more detached, gentlemanly view of the estate. In the Red Book, Repton chided those who coveted land. He was glad that the Sheringham estate was bounded by the sea and by property which Upcher could not purchase. The vain illusion of making everything in sight his own was thus impossible. Upcher was actually intent on enlarging his estate but wished also to enlarge his hospitality. He and Repton agreed that 'exclusive possession is absolute folly'. The Red Book describes plans to extend the hospitality of the estate to wealthy tourists and the local poor.

Repton designed a rotunda, positioned about three-quarters of a mile from the house, where tourists could take their refreshments and enjoy a view of the house set in its park. In turn, these visitors, like the harvesters in the cornfield, would enliven the scenery for the family and their guests in the house. For visitors the rotunda was essentially a useful building, for the residents it was essentially an ornamental object.

Repton devoted considerable space in the Red Book to the treatment of the local poor. He wondered why on some estates they were prepared to rise at night to fight for their squire and on others would rise at night only to steal his game. He concluded that where squires and farmers were detached from labourers, the poor were both oppressed and rebellious. On those estates whose owners detested the sight of men, he had seen lame and blind beggars driven from the door. Where farmers no longer worked alongside their employees but watched over them, labourers were, when past work, turned over to the parish officer and incarcerated in 'prisons erected under the name of workhouses'.[48] The workhouse at Sheringham, conspicuously positioned near the entrance of Upcher's estate, was a distasteful reminder of these developments. Repton suggested how it might be made to appear a more benevolent institution:

> The workhouse instead of being an object of disgust to the rich and terror to the Poor, might be made to look more like a hospital or asylum and less like a Prison; by removing the high wall the street might be converted into a neat *Village green* with its benches and a *May-pole,* that almost forgotten emblem of rural happiness and festivity.[49]

Upcher wanted a main entrance to his estate that could be reached without passing through a village that Repton thought was woefully lacking in elegance, comfort or gentility. The new house was to be reached not through the farm but along a winding drive through the woods. It was much further from the village than the house Cook Flower inhabited, but Repton was confident it would not be disconnected from the village. In the absence of a village green, Repton suggested the beach was a suitable arena for promoting greater fellowship and for resolving an issue of intense conflict hunting game. Abbot Upcher's favourite pastime was coursing. Unlike shooting, which involved only a family and their guests and which Repton described as a 'selfish enjoyment',[50] coursing was a 'social enjoyment' which could involve the whole community in public matches held on the sands at low tide.

> This promotes a mutual intercourse betwixt the Landlord, the Tenant and the Labourer, which is kept up at little expense and secures the ready and reciprocal assistance of each to the other. This is the happy medium between Licentious Equality and Oppressive Tyranny.[51]

In a coastal region exposed to the threat of a French invasion such fellowship was felt to be essential. Repton thought his advice on growing trees to withstand gales would be a lesson to the moralist as well as the planter: 'England's combined Oaks resist the Sea,' he declared, 'Emblem of Strength, increas'd by Unity'.[52]

The moral landscape that Repton envisaged at Sheringham was meticulously planned. Some parts, like the oak plantations, were purely emblematic. Other parts had a functional role in the improvement of social relations. It is remarkable how calculated this improvement was. The Upchers were protected from spontaneous social encounters by the siting and design of their house and the layout and regime of the park. The visits of both tourists and the poor were strictly controlled. One day in the week respectable looking tourists could sign a book at the lodge and make walks within a limited area well away from the house. At stated periods, the destitute could be admitted to the house to receive milk and leftovers. Repton designed the house to ensure that such calls need not disturb the family. Villagers could gather dead wood one day a month while being supervised by the keeper. This was granted as much to prevent the stealing of wood and disfigurement of trees as to increase the comfort of the poor. It was disorder as much as deprivation that alarmed Repton. At a time of Luddism in the industrial districts, Repton was relieved that there were no factories nearby whose workers might ferment sedition and fan the flames of discontent. The Red Book for Sheringham is a brief not so much for the exercise as for the display of paternalism. E. P. Thompson notes that landed paternalism in Georgian England was characterised more by gestures and postures than real day-to-day responsibilities, by:

> occasional dramatic interventions: the roasted ox, the prizes offered for some race or sport, the liberal donation to charity in times of dearth, the

application for mercy, the proclamation against forestallers. It is as if the illusion of paternalism was too fragile to be risked by more sustained exposure.[53]

The Upchers carried out their brief with a great deal of deliberation. They moved to Sheringham in October 1812 and into the farmhouse that Cook Flower vacated. They laid the foundation stone for the house a year later with a prayer that its door be open to those in distress and its stores unlocked to feed and clothe the destitute. In the severe winter of 1813–14, Upcher distributed food and fuel to the poor. On Christmas Eve 1813 he gave meat to 72 poor families (about half the population), and on Christmas Day ceremoniously had the parish poorhouse pulled down. Upcher supervised the changes to his estate. In November 1815 he wrote in his diary: 'the mornings I have dedicated to external improvements in my woods and fields and the evenings to sacred mental acquirements'. In January 1817 the new house, the pivot of Repton's moral landscape, was built and ready to be furnished. The Upchers prepared to move in that summer but never did so. Abbot Upcher contracted meningitis and died in the old farm in February 1819 aged only 35.[54] 'Had he lived longer', wrote his grand-daughter, 'he would indeed have been remembered as the *first good squire* of Sheringham'.[55] Charlotte Upcher outlived her husband by 38 years and in that time carried on the stewardship of the estate. She established a Friendly Society, allotment scheme, and adult school, loaned money to fishermen, and provided the village with a lifeboat. Among her more notable acts were saving a local fisherman from a press gang and securing a pardon for a woman condemned to be hanged for stealing.[56] She remained living in the farmhouse and her son was to become the first tenant of Sheringham Hall upon his marriage in 1838.

Hare Street

Repton intended to promote a sense of community at Sheringham and, in his phrase 'humanise' the scenery of the estate. In the Sheringham Red Book, the ideal of social propinquity contrasts with the actual distance of villagers from the Upchers' new house. This paradoxical way of seeing, which John Barrell identifies in contemporary works by Constable and Wordsworth,[57] is pronounced in Repton's design for his own garden at Hare Street in Essex. The description and illustration of the improvements he made to his garden was the last work Repton wrote for publication and was intended to sum-marise his own outlook (Fig. 7.4).[58]

Repton was concerned about the arrival in Hare Street of war profiteers, one of whom had driven out villagers by doubling rents.[59] Repton intended his improvements to produce a more humane outlook. He explained that the road was originally five yards from his window and that he obtained

Figure 7.4 Repton's improvements at Hare Street.

permission to extend his fence, taking in an area customarily used to graze droves of cattle, pigs and geese:

> By the *appropriation* of twenty five yards of garden I have obtained a frame for my landscape; the frame is composed of flowering shrubs and evergreens; beyond which are seen the cheerful village, the high road, and that constant moving scene which I would not exchange for any of the lonely parks I have improved for others . . . I love to see mankind.

The illustration of the improvements reveal that Repton preferred to observe tokens of mankind in the distance rather than encounter men at closer quarters. The improvements appropriated the village as a view by erasing its salient social character. The passing coach, the butcher's shop, the gaggle of geese, the beggar who leans cap-in-hand on Repton's fence, are excised. The village becomes less active and involving; more a landscape, less a place. Repton does not comment on the beggar, lame and blind like those who haunt the Sheringham Red Book, and probably one of the war veterans who were swelling the ranks of the unemployed. Following the argument in the Sheringham Red Book, the implication may be that he has been relieved, not uncharitably dismissed. Nevertheless the improvements disclose that what really matters is not that the village *is* cheerful but that it *looks* cheerful; in other words, what is at issue is not so much the cheerfulness of the villagers as the cheerfulness of Repton, the observer. The improvements thus represent an ultimately complacent view of rural life.

Conclusion

This essay has concentrated on the relationship between the design of a Georgian landscape garden and the world beyond its perimeter. Landscape gardens, both those that were built and those that remained literary or artistic constructions, were instruments for interpreting and changing this world. Landscape composition involved selectively and strategically manipulating the rural world – emphasising some aspects, disguising or concealing others, separating what was dialectically linked, harmonising what was antagonistic.[60] The values that informed the design of landscape parks were many, various and conflicting. Humphry Repton incorporated and modified conservative landscape conventions which emphasised the moral obligations of landowners, which were critical of the prevailing social order and nostalgic. Repton's mature rural vision was both active and complacent. His landscapes imply intervening in the rural world but are essentially attempts to dissolve its tensions and disturbances into an agreeable image. I reconstructed this process at Sheringham by examining the relationship between Repton's design, the existing landscape, the values of Repton and his client and the social and economic constitution of the area at the time of the commission.

The central issue of the Sheringham commission, and of much Georgian rural art and literature, is the place of the landed gentry in an agrarian society. Victorian moralists shifted their attention away from the countryside to the industrial towns and cities and to the role of industrial capitalists. They saw a factory town as more representative of the condition of England than a landed estate. The characteristic moral landscapes that Victorians envisaged and built were intended for factory workers and mill owners, not farm workers and squires.[61]

Acknowledgements

I wish to thank Mr Thomas Upcher for his hospitality at Sheringham Hall and his permission to reproduce the illustration from the Red Book for Sheringham. John Barrell and Hugh Prince read an earlier draft of this essay and I thank them for their helpful criticism.

Notes

1 Watt, I. 1968. Two historical aspects of the Augustan tradition. In *Studies in the Eighteenth Century*, R. F. Brissenden (ed.), 67–89. Canberra: Australian National University Press.

2 Duckworth, A. M. 1971. *The Improvement of the Estate: A Study of Jane Austen's Novels*, 36–60. Baltimore: Johns Hopkins Press.

3 Everett, N. H. 1977. *Country Justice: The Literature of Landscape Improvement and English Conservatism with particular reference to the 1790's*. Unpublished Ph.D. thesis, University of Cambridge.

4 Cowper, W. 1785. *The Task*, Book III, lines 755–6.

5 Batey, M. 1974. Oliver Goldsmith: an indictment of landscape gardening. In *Furor Hortensis*, P. Willis (ed.), 57–71. Edinburgh: Elysium Press.

6 George, I. c.22, The Black Act. Reproduced in Thompson, E. P. 1977. *Whigs and Hunters: The Origin of the Black Act*, 270–2. London: Penguin.
See also D. Hay, P. Linebaugh, J. G. Rule, E. P. Thompson and C. Winslow (eds) 1977. *Albion's Fatal Tree: Crime and Society in Eighteenth Century England*, 270–7. London: Penguin.

7 The Vicar of Wakefield, published in 1766. Reprinted in *The Complete Works of Oliver Goldsmith*, 1886, 690. London: Routledge & Kegan Paul.

8 Williams, R. 1973. *The Country and the City*, 120–6. London: Chatto & Windus.

9 Mack, M. 1969. *The Garden and the City: Retirement and Politics in the Later Poetry of Pope*, 53–4, 91–4. Toronto: University of Toronto Press.
Williams, R. op. cit., 58–9.

10 Gilpin, W. 1786. *Observations on the Mountains and Lakes of Cumberland and Westmoreland*. Vol. 2, 44–5. London.
Gilpin, W. 1791. *Remarks on Forest Scenery*. Volume 2. 166–7. London.

11 Everett, N. H. op. cit., 135–9.

12 Price, U. 1795. *Essays on the Picturesque*, Vol. 3. 159–60.

13 Owen, D. 1965. *English Philanthropy 1660–1960*, 99–100. London: Oxford University Press.

14 Price, U. 1797. *Thoughts on the Defence of Property*, 20, 28.
15 The definitive biography is Stroud, D. 1962. *Humphry Repton*. London: Country Life.
16 Quoted in the 'Memoir' in Loudon, J. C. (ed.) 1840. *The Landscape Gardening and Landscape Architecture of the Late Humphry Repton*, 10.
17 Stroud, D. op. cit., 28.
18 Mason, W. 1771–9. *The English Garden*, Book 2, lines 21–2.
19 Whateley, T. 1771. *Observations on Modern Gardening*, 144–5.
20 Gerardin, R. L. 1783. *An Essay on Landscape*, 150.
21 Stroud, D. op. cit., 28, 35.
22 Malins, E. 1976. *The Red Books of Humphry Repton*, 12–13. London: Basilisk Press.
23 Repton, H. 1794. *Sketches and Hints on Landscape Gardening*, 78–81.
24 Repton, H. 1794. *A Letter to Uvedale Price Esq.*, 5.
25 *Red Book for Attingham*, prepared for Lord Berwick in 1798, no pagination, reprinted by the Basilisk Press, London, in 1976. The definition of 'landscape' is taken from Samuel Johnson's *Dictionary*. The relevant work by Edmund Burke, which Repton quotes, is the essay 'On Taste' which forms the introduction to the 1793 edition of Burke's *Philosophical Enquiry into the Origin of our Ideas of the Sublime and the Beautiful*.
26 *Red Book for Attingham;* Repton, H. 1803. Observations on the theory and practice of landscape gardening, 206–7; *Red Book for Armley*, prepared in 1810 for Benjamin Gott, no pagination, in the private collection of Paul Melon at Upperville, Virginia.
27 Repton, H. 1794. op. cit., 81.
28 Repton, H. 1803. op. cit., 92–7.
29 Repton, H. 1806. *An Enquiry into the Changes of Taste in Landscape Gardening*, 56, 105.
30 Repton, H. 1803. op. cit., 138.
31 Repton, H. 1788. *Variety* **58,** 61–2.
32 Repton, H. 1816. *Fragments on the Theory and Practice of Landscape Gardening*, 233–4, 192–3.
33 *The Red Book for Sheringham* is owned by Thomas Upcher of Sheringham Hall and is on loan to the Royal Institute of British Architects. It was reprinted by the Basilisk Press, London, in 1976 in a boxed set with the Red Books for Anthony and Attingham and an introduction by Edward Malins.
34 *History and Antiquities of the County of Norfolk*, Vol. 3, 1781, p. 101.
35 Pratt, S. J. 1801. *Gleanings in England*, Vol. 1, 431, 446.
36 Bartell, E. Jr., 1800. *Observations upon the Town of Cromer and the Picturesque Scenery of its Neighbourhood*, 75–9. London. This was republished in an enlarged edition in 1806. Bartell was a member of the Norwich Society of Artists and helped popularise the area with other members. See Hemingway, A. 1979. *The Norwich School of Painters*, 29. Oxford: Phaidon.
37 Bartell, E. Jr. 1804. *Hints for Picturesque Improvements in Ornamental Cottages and their Scenery*. Here Bartell refers to the alarm Repton expressed in *Observations* at the 'expanded avarice' of farmers. He also quotes at length from Nathaniel Kent, an experienced land agent and influential reporter on the agriculture of Norfolk.
    See Kent, N. 1794. *Hints to Gentlemen of Landed Property*, 45–7 and in the 1796 edition, 156, 172. In the 1770s Kent organised and managed the enclosure and emparking of Felbrigg on principles of utility and benevolence. Repton frequently visited the Felbrigg estate when he resided in the neighbouring parish of Sustead. He knew and admired Kent and was no doubt influenced by his writings and practice.

See Ketton Cremer, R. W. 1976. *Felbrigg: The Story of a House,* 172–6. Ipswich: Boydell Press.

Stroud, D. op. cit., 22, 161.

38 The information in this paragraph is taken from Mason, R. H. 1884. *The History of Norfolk,* 454–85. London.

Peacock, A. J. 1965. *Bread or Blood: A Study of the Agrarian Riots in East Anglia in 1816,* 11–38. London: Gollancz.

Taylor, R. 1970. *The Development of the Old Poor Law in Norfolk.* Unpublished Master's Thesis, University of East Anglia.

Hobsbawm, E. J. and G. Rude 1973. *Captain Swing,* 18–33. London: Penguin.

Digby, A. 1978. *Pauper Palaces.* London: Routledge and Kegan Paul.

39 The Sheringham Enclosure Award is deposited in the Legal Department of Norfolk County Hall, Norwich. The Land Tax Assessments for Sheringham are deposited in the Norfolk County Record Office, Central Library, Norwich.

40 Carter, M. J. 1980. *Peasants and Poachers: A Study of Rural Disorder in Norfolk,* 2. Woodbridge, Suffolk: Boydell Press.

41 Errol, A. C. 1970. *A History of the Parishes of Sheringham and Beeston Regis,* 85. Norwich: privately printed and published.

42 John Barrell makes the same general point about literary convention and empirical observation in his reading of John Clare's poem *The Parish:* see Barrell, J. 1972. *The Idea of Landscape and the Sense of Place 1730–1840: An Approach to the Poetry of John Clare,* 195–8. Cambridge: Cambridge University Press.

43 Pigott, E. (n.d.) *Memoir of the Honourable Mrs Upcher of Sheringham,* 20–1, 36, 44. No place of publication or publisher. Emma Pigott was a daughter of Abbot and Charlotte Upcher.

44 The image of gold was often used to signify Whiggish corruption. For Pope's use of the imagery of gold and paper money see Mack, M. op. cit., 87–8. The Sheringham Red Book is not paginated. Unless otherwise indicated all subsequent commentary and quotations from Repton are from the Red Book.

45 Pigott, E. op. cit., 69.

46 Repton here quotes the following lines (268–72) from a satirical poem *The Landscape,* written by Richard Payne Knight and first published in 1794.

> Now, not one moving object must appear
> Except the Owner's Bullocks, Sheep and Deer
> As if his Landscape was made to eat
> And yet he shudders at a crop of wheat.

Ironically the poem also contains an attack on Repton, coupling him with Capability Brown.

47 Pigott, E. op. cit., 35.

48 The liking of workhouses to prisons was commonplace but the use of this image by George Crabbe to striking effect in Letter 18 of *The Borough* (1810), where it refers to an incorporated East Anglian workhouse, may well have influenced Repton's apprehension of the Sheringham workhouse.

49 Shortly after the Sheringham commission Repton was asked by his brother Edward, vicar of Crayford in Kent, to design a workhouse for the parish. This is illustrated and described in *Fragments,* 229–30. With its mullioned, dormer windows it evoked the hospitality of old almshouses. It was built with a garden and terrace from which the inmates could enjoy the view towards Canterbury and the coast. Repton noted ruefully that the parish elders decided the view made the site too profitable for paupers to enjoy at the ratepayers' expense and sold it for building private houses.

50 Cf. Goldsmith in *The Deserted Village* (Note 5):

> 'His seat, where solitary sports are seen,
> Indignant spurns the village from the green.'

51 Repton used a similar political image in his *Letter to Uvedale Price* where he likened the English landscape garden, a 'happy medium' between wild nature and the imposing formality of French gardens, to the English constitution, 'a happy medium betwixt the liberty of savages and the restraints of despotic government'.

52 In 1803 a convoy of merchant ships off the Sheringham coast was mistaken for a French invasion fleet and in 1814–15 troops were stationed along the cliffs to repel another feared French invasion. See Brooks, P. 1980. *Sheringham: The Story of a Town*, 7. Cromer: Poppyland.

53 Thompson, E. P. 1974. Patrician society, plebeian culture. *J. Social History* **7**, 383–405 (390).

54 Pigott, E. op. cit., 55–6, 61–4, 70.

55 Pigott, B. 1890. *Recollections of Our Mother Emma Pigott,* 4. London: Nisbet.

56 Upcher, C. 1932–3. Collections and recollections. *Sheringham Parish Magazine.* My thanks to Campbell Errol for transcribing the passages on Charlotte Upcher.

57 Barrell, J. 1980. *The Dark Side of the Landscape: The Rural Poor in English Painting 1730–1840*, 140–1. Cambridge: Cambridge University Press.

58 Repton. H. 1816. op. cit., 235–6.

59 Letter from Humphry Repton to Sir Harry Fetherstonehaugh, 20 October, 1816, quoted on p. 83 of Meade-Fetherstonehaugh, M. and O. Warner 1964. *Uppark and its People*. London: George Allen & Unwin.

60 For a suggestive theoretical discussion of landscape conventions and ideology see Turner J. 1979. *The Politics of Landscape: Rural Scenery and Society in English Poetry 1630–1660*, 186–95. Oxford: Blackwell.

61 Daniels, S. J. 1980. *Moral Order and the Industrial Environment in the Woollen Textile Districts of West Yorkshire, 1780–1880*, 118–290. Unpublished Ph.D. thesis, University of London.

8 Places, conservation and the care of streets in Hartlepool

KATHERINE A. OLIVER

'When you know how to look, you can discover the spirit of an age and
the physiognomy of a king even in a door-knocker.'

Victor Hugo *Notre Dame de Paris*

More people live in streets and along roads than ever before. More money is
spent on destroying some streets and conserving others than at any time in
the past. Simultaneously, more people are impoverished by these changes,
by their loss of place in society and by their loss of places in their own small
worlds as the world outside their own responsibilities grows bigger and more
hostile. To restore confidence and to enjoy streets, we must recognise our
own predicament. Understanding and encouraging the ways in which places
are created, maintained and enjoyed demands an awareness of the relations
between people, places, ideals and ideologies. For example, the current
interest in the conservation of the built environment, the degree of controversy which surrounds the subject, and the variety of beliefs and activities
known as conservation suggest that we should question the premises about
our care of streets. The adherents to the conservation movement are attempting to change the values of mid-20th century Western Europe and to provide
a working paradigm for the realisation of the ideals of conservation in the
built environment.[1] Justification for conserving the built environment has
relied upon analogy and the easy rhetoric of capitalist economics, and
most 20th-century conservation policies tend to favour those places possessing a heritage which evokes past experience. In this case, changes to the
built environment are in danger of expressing only one ideology which may
be meaningless for the majority of people. The assumptions of both the
designers and the users of places must be announced, because they form part
of the totality of intentions which materialise in streets and because their
assumptions may not be wholly appropriate for the purpose of changing
streets.

I propose that the opportunity for people to establish the identity of places
for themselves is being threatened not encouraged, by the practice of *conservation,* as exemplified in the *care of streets.* Further, this practice of conservation
has provoked and perpetuated an *imagery* in streets – a collective view of a set
of physical features – which obscures the political and economic forces by

which it is created. These forces, such as the select designation of Conservation Areas, the corporate power of multiple retail outlets, and the separation
of the roles by which people create and alter places, in themselves defeat the
aims of conservation. Trees are planted in housing areas and then uprooted;
paving is laid one week and dug up the next; seats are provided which are
either too uncomfortable to sit on or are put where no-one wishes to sit. All
are apparently trivial examples, but they are ubiquitous and costly and they
suggest contradictions in social aspirations, in government policy and in
planning.

In this chapter I discuss the care of streets in Hartlepool with two main
aims: first to illustrate the conflict between popular experience of streets and
political control over them, and secondly to demonstrate whether or not the
current ideology of conservation is in danger of stereotyping ideas and
physical forms ('motifs'), which threaten the identity of places. The research
seeks to establish how clearly streets represent the intentions of the people
they serve and by whom they are created, and whether their appearance is
acceptable only temporarily in accordance with certain beliefs prevailing at
the time. In particular, the following questions are posed. What is the present
appearance of the streets and what are the motives behind the changes which
occur? What are people's feelings towards the streets? What are the disparities
between the ideals of conservation and the present appearance of streets?

The first part of the chapter contains a summary of a street survey that I
carried out, which establishes the present appearance of four main streets in
central Hartlepool.[2] The second part presents documentary material and
information from interviews as a commentary upon the process and effects of
changes to the street since the 1950s, and upon the agents responsible for that
change. The third section uses this view of the change and conservation of
streets as the basis for appraising conservation in Hartlepool and, in particular, for identifying those ambiguities which tend to undermine conservation. Finally, I shall discuss the wider implications of the care of streets in
Hartlepool for making and conserving places.

The street survey: the motifs of conservation

Despite the fact that the 'history of West Hartlepool . . . is of American
brevity',[3] the town centre has migrated twice since its foundation in the
1830s. In 1835 a railway was constructed from the Durham coalfields to old
Hartlepool harbour. The railway company did not own the docks and
quarrels arose over the dues for coal shipment. As a result, the railway's
entrepreneur, Ralph Ward Jackson, decided to build a new harbour at the
adjacent hamlet of New Stranton. Jackson himself laid out the new settlement known as 'West Hartlepool' on a grid pattern, with streets named after
royalty, local landowners and other ports. By the end of the century, Church

Street had succeeded Albert Terrace as the centre of the town. However, the vicissitudes of heavy industry, the rapid increase in population,[4] the impact of the world recession between the wars, and the uncontrolled pattern of land use in the town centre all provoked the need for town centre redevelopment. In 1937, the civic authority commissioned Hartlepool's Regional Town Planning Scheme. Work on this was interrupted by World War II and war damage exacerbated the problem.

Immediately after the war in 1946, Max Lock, an eminent town planner, was appointed to prepare a comprehensive improvement plan. He advocated wholesale reconstruction of large areas of bad development, and zoning to separate out the various different functions of the town that had become muddled together in the urban fabric. In particular, he argued for the re-development of the town centre. The numerous, unprosperous street shops were to be regrouped into a new shopping centre providing twice the existing shop frontage. He proposed a new civic centre to include a Town Hall, library, art gallery, two theatres, hotels and a central park. The construction of this new centre and its 13 associated car parks would require the demolition of 2544 houses in 41 streets, drastically reducing the population of the central area.

This plan formed the basis for all subsequent policies over the next 20 years. In 1947, the Borough Council endorsed Lock's proposals in principle, and transposed most of them into West Hartlepool's Development Plan, which was approved by the Minister of Housing and Local Government in 1951. The scheme, however, was to become the focus of an extensive party political battle. Eventually, the existing 11 discrete units with corner buildings and shops were demolished and became one massive site for the Middleton Grange Shopping Centre, which was finally opened in 1971. The cultural facilities envisaged by Lock are still at the 'planning stage' 30 years later, and a leisure centre – a facility now common in many towns that are much smaller than Hartlepool – has yet to be started. Against this, it should be noted that considerable advances have been made in separating out different land uses, lighting has been much improved, trees have been planted, and street adver-tising is no longer so prolific.

The values inherent in these plans and the ensuing changes made are examined in the four streets that I surveyed (see Fig. 8.1). I selected these four because they are complementary, distinctive, and all have a claim to the status of 'high street' in a town which lacks a 'High Street' as such. *Church Street* and *Church Square* attest to bygone grandeur but not to centrality or popularity. *Victoria Road* is central and has the most community services, but is not the main shopping street. *York Road* has many of the characteristics of a 'high street', but it is neither dominant nor focal. The near by Middleton Grange Shopping Centre is not a street in the strict sense, nor is it complete, yet clearly it qualifies for inclusion as the commercial centre.

For each of these streets, I established their present appearance by enumer-ating and describing every feature which caught my eye and imagination,

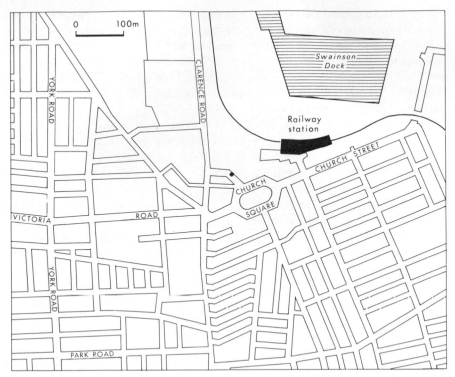

Figure 8.1   Location of streets surveyed in Hartlepool.

including the motifs of conservation. These are the presence and combination of such features as patterned paving, semi-mature trees, seats, bollards, litter bins and plant bowls, the 'painting out' of pipes and wires on buildings, and the control of signs and advertisements. I made repeated visits to these streets at various times of day and year to acquire a graphic and photographic record of the forms and colours, patterns and activities found in these streets. These are summarised in 'catalogue plans', an example of which is shown for York Road in Figure 8.2.

The surveys are represented here by examining in detail the results for one street. *York Road* is bounded by red-brick buildings with wood and stone details, pitched roofs and added shop fronts. The building line is broken on the east by an extensive car park. Here, the distant horizon is enclosed by a high wall of uniform orange brick and bare concrete. The wide road and low roof-line give an impression of calm that is untypical of a high street, enhanced by the sedate progress of cars and by the many pedestrians standing at bus stops. Considering its central position, commercial function, main road status and unexceptional architecture, York Road is a comfortable street without being 'touristy' or '*chic*'.

Yet in terms of Cullen's attributes of places,[5] York Road is meagrely

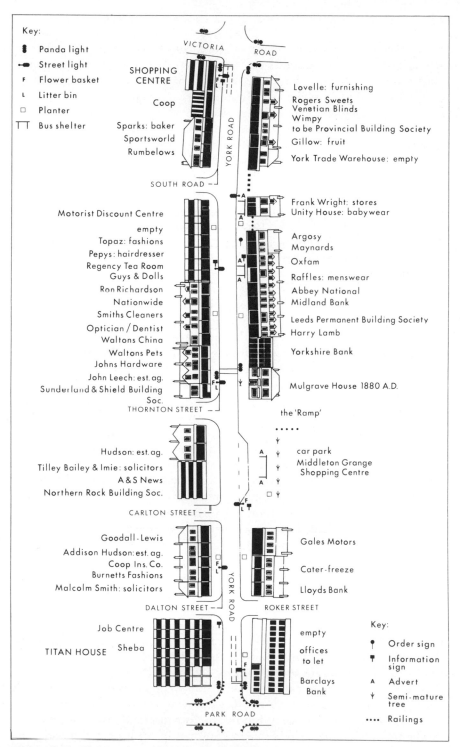

Figure 8.2 The Catalogue Plan for York Road.

bequeathed. No significant changes in level command a view. There are no strong focal points, spotlights or 'handsome gestures' – other than a fine clock over Harry Lamb's jewellery shop. No narrows or incidents, no sense of infinity or mystery make York Road a place. No one has invested time or money to provide any buildings of noticeable beauty or architectural distinction. Where there has been modern infill, designers have matched some features with existing ones but these are negated by other incompatible elements. The designer of the Yorkshire Bank matched the building height and eaves line, but the roof is flat and the glass is black in brown metal window frames. Titan House and Northern Rock House are striking buildings interposed into a street of otherwise unostentatious and solid red-brick houses. The architects of Northern Rock House chose white weatherboarding, and Titan House, a six storey building with concrete cladding, also violates the traditions of material and size. The latter also challenges the adjacent church, darkens the entrance to Park Road and diminishes the impact of the church tower.

Paradoxically, it is these unpleasing eyecatchers that draw attention to features that, in other circumstances, might have been taken for granted even though they contribute to the overall impression of the street. The terraces of houses built before the war have been adapted individually. Dormer windows, brickwork painted and rendered, the preservation of some bay windows on ground and first floors, the chimney stacks and downpipes, combine in a view that is appealing in its complexity rather than its purity. In turn, these often shabby features show to advantage the prosperous neatness of new buildings, and to disadvantage their monotonous facades lacking functional details, stone, or patterned and coloured brick.

Street furniture and landscaping indicating planned, concerted change by the local authority contrast with the incrementally developed buildings which line the road. The car park is recognisable as such, even when empty, by the lines, arrows and instructions painted in white on the tarmac. The semi-mature trees planted on the perimeter are still too spindly to 'soften' its appearance. Concrete pyramidal tiles separate pavement from car park, pedestrians from cars. A double row of trees confirms this barrier. Cobbled slopes outline the ramp up and across the car park, and white railings prevent pedestrians from falling off the edge. The road is divided into car and pedestrian space by a step, which is itself marked by the kerbstone and the change from tarmac to concrete and paving slabs. The highway engineers must fear that these signs are inadequate for pedestrian safety, for there are three stretches where railings reinforce the kerb. The instructions of signs are reinforced in two locations by bollards that block the way for cars. Hence there are no less than four different indicators of this segregation in road use and space. These 'markers' are in what Cullen termed the 'functional tradition' of design,[6] which it seems is now largely a local authority tradition.

The street furniture is predominantly post-war, made of steel, plastic and glass, in one or more of five colours, according to their purpose and age. Benches have been provided, but are somewhat derisory in number (two) and design (they have no arms or backs). One is situated by the telephone kiosk near the junction with Victoria Road. The other bench located near to the intersection with Park Road, neither serves a bus stop nor an obvious meeting place; it would only seem to provide a view of the cars stopping at the traffic lights. Tall, monotonous, green metal lamp posts stand at the kerb on both sides of the road. The lanterns' orientation demonstrates their allegiance to car-drivers, not to pedestrians. The lamp posts themselves do not define a space; rather they belittle the space described by the buildings and by the trees, which are perhaps only half their height. The flower baskets which cling to them are disconcertingly placed just under the half-way mark of the height of the posts, seemingly emphasising their 'over-growth'. Another reminder of the banishment of natural features from York Road are the planters, referred to locally as 'sleeping coffins', that lie strategically at intervals in the middle of the pavement. Although more noticeable for their solidity and location than for their beauty or contents, they are well-tended and colourful when the plants are in flower. Street advertisements, part of the rhetoric of commerce, are to be found exclusively on Council property — small ones on the yellow plastic and brown metal litter bins that adorn the lamp posts, larger ones on the bus shelters.

The process and agents of change

The street catalogue for York Road contains little that is remarkable in itself, but it does provide a visible and tangible record of a complex local history which bears the impress of the general history of Hartlepool outlined earlier, and of the particular care of York Road. For example, there is tangible evidence of the changes in the nature of the street which began around the turn of the century. York Road served initially as a route from old Hartlepool southwards to Seaton Carew. Gradually, traders opened shops to serve people living in the new houses built in the western part of the rapidly growing town. Since York Road was not originally in the town centre, there is no lavish architecture from this period in the street. Owners did not feel the 'civic pride' necessary to maintain an image in what was then a peripheral area. Nonetheless, in 1902 traders paid for 55 trees to be planted, each with an enclosing guard.[7] By 1928, they had gained the confidence to promote York Road as the town's major shopping street, and succeeded in attracting banks and commercial enterprises (which now constitute the dominant land use). Local initiatives in caring for the street have continued. In the 1960s, the traders initiated two improvement schemes to brighten up the street. The first, in 1962, included the installation of the stone planters and hanging baskets; the second, in 1967, aimed to offset the competition from traders in the nascent Middleton Grange Shopping Centre.

Only part of the explanation for the process of change can be found in local initiatives. The wider context of official planning, and especially of the conservation policies, needs to be examined. Since the advent of the planning system in Britain, the town council has assumed increasing control over Hartlepool's aesthetic. The role of Cleveland County Council with respect to the care of streets in central Hartlepool is to advise rather than to direct action or plan implementation.[8]

Opinion in the town varies as to how well Hartlepool Borough Council has discharged its responsibilities. The general feeling is sceptical, with criticism of both the goals of policy and the methods of implementation. For example, a typical view is that: *'On the Hartlepool Borough Council, no one has any feel for what the town has got. Its main concern is dog muck on the pavements'*. A Cleveland County Council officer privately advanced the view that the Borough Council is 'behindhand' with conservation policy. In 1981, the Council has still to publish plans for the five Conservation Areas already designated. Moreover, while there are signs that the Council has started to rehabilitate terraced houses and to encourage people to move back into the centre of town, there was a common feeling that the Council is bent on demolition, whatever the weight of public opinion or the rationale for conservation. It should be pointed out, however, that the Council's policy in this matter does not involve party political conflict. Most councillors agree that there are more pressing needs on which to spend the rates than conserving historic buildings.

While councillors are often in accord, the departmental separation of the Council officers' responsibilities is debilitating. Indeed, I found it difficult to trace which Council officer is responsible for which items of street furniture. Although Cleveland County Council has authority over the main roads, the Borough Council acts as its agent on everyday matters of street furniture. The design of signs is dictated by Department of Transport regulations; their siting is a matter for the Engineers. The latter can also decide on the design and location of bollards, litter bins, street lights and bus shelters. Most seats are paid for by private donation. When they are provided by the Council, sites are chosen by a sub-committee of the Borough Council in collaboration with the Parks Department. The same partnership deals with trees.

On inquiring about the various motifs of conservation, litter bins and street advertisements elicited the strongest public reaction. Litter bins are a 'sore point': *'we can put 10 000 bins in Hartlepool and we would still have dirty streets unless there is effective education'*. One councillor did feel that the provision of more bins and the anti-litter campaign were improving the situation, but felt that the shortage of staff to empty the bins coupled with the inadequate design and siting of the bins negates this achievement.

There are a variety of views on how to control street advertising, but there is a consensus that the Council should exercise some control *'to guard against poor taste'* and *'immoral advertising'*. All agreed that rigid rules are undesirable, and that the Council should seek to find *'a middle ground which will enhance the advertisements and the environment'*. Particular concern was expressed about fly-posting, but some made the distinction between unauthorised fly-posting on walls and buildings and the more acceptable bill-posting inside shop windows. People were less decisive about the control of fascias and shop signs, either feeling that there was no need for control or that it should merely *'prevent them running riot'*. Those against control cited the uniformity of the Shopping Centre, which *'needs protruding signs, more latitude and brightness because everything the same height hardly adds to the creation of colour and character'*.

Indeed it is instructive to consider briefly the experience of the Middleton Grange Shopping Centre. Here the land and property freehold was retained by the local authority, and international firms of architects and property managers were engaged as developers.[9] Accordingly, universal motifs of modern design are evident in the Centre. The vernacular traditions of red brick and low skyline are conceded, but the scale and style of the buildings, the lack of small spaces to serve as meeting-places, dark concourses and absence of small details violate these traditions. Moreover, the homogenised features of the Shopping Centre combined with a general lack of the patina of age and life gives the development more than its fair share of stereotyped features. In the Centre, the consultant architects co-ordinated the designs both for the individual shop fronts and for street furniture. All advertisements, fascias and lettering are controlled by the Council and strict policies are adhered to. Although there is no specific colour scheme, the Council's Corporate Estates Committee ensures that 'no conflict exists with respect to colour and design with those structures and furniture originally supplied'.[10] Until now, explicit policy has been to keep the facades relentlessly uniform in this commercial environment but, in future, projecting signs indicating trade will be permitted.

Reactions to the Centre are mixed. The Council states that the 'image of Hartlepool as a go-ahead town has been enhanced'.[11] The Civic Trust bestowed an award upon the Centre's architects in 1972 and, not wholly surprisingly, the Cement and Concrete Association commended the board-marked treatment of the exposed concrete.[12] The local newspaper[13] commented upon the 'unusual and attractive roof design of the covered market' (a feature I did not notice – do others?), and continues to relay information about and reactions to the Centre, a decade after it was officially opened. This is perhaps indicative of the political visibility and social symbolism of this new form of 'high street'.

The shortcomings of the Centre are well aired. Although the Council describes the Shopping Centre as 'linking naturally with the older more established shopping areas',[14] the letting agents observe that 'concern is shown in some quarters that covered schemes are incompatible with established shopping patterns, particularly in historic centres, and the debate on covered versus open designs is likely to continue'.[15] Criticisms range from *'very conventional'* to a *'concrete monstrosity with appalling design, appearance and function'*. Particular shortcomings include the limited mixture of firms that have been attracted to the Centre, and its design. Disabled people in wheelchairs cannot reach the top deck unless helped up the ramp. The darkness of the roofs and lack of shelter for the few seats that are available defeat the purpose of the malls. On the upper deck, the wide concourses are bleak and under-endowed with refuges or seats on which to wait, eat or watch. In winter it is often colder in the Centre than outside, with the malls acting as wind tunnels.[16] Widespread dislike is expressed about the building materials used in the Centre and about its littered state, despite continual vacuum cleaning. In general, the Centre is seen as *'too open, big and chilly'*, a useful facility rather than a pleasing place.

In assessing the Centre, planners and architects see physical form as an active constituent in social change. On the other hand, people who I interviewed identify with the function of the place and rarely with its physical attributes. Perhaps, for most people, physical form is a subconscious condition influencing their ability to create places rather than a necessary, identified, formal context. The local authority and the letting agents have capitalised on this by fostering an image of a profitable regional centre, relying on abstract notions of land use, planning and economics. As controllers of the capital invested and design adopted, they are in a position to ignore potential drawbacks. Their design criteria fail to accommodate all the values that are already expressed in the surrounding streets. They have chosen to build a modernist emblem of civic pride and progress which has been appliquéd to the existing fabric of the town. The results may be seen in the barren eastern side on Stockton Street and in the unsympathetic impact of the huge car park and truncated side-streets on York Road. The Centre does not reconcile the interest of developers, builders, council officers, councillors, leaseholders, traders and shoppers, but merely overrides them. In the absence of a traders' association or of concern by the Civic Society, there is no adequate pressure to mitigate the Council's control over many aspects of the Centre's aesthetic. Neither this control nor the capital invested can substitute for the trader's care for his property and the streets.

The ambiguity in streets

A radical and serious paradox does not hang upon a removable confusion, but is demanded by the complexity and inherent ambiguity of what is being expressed.[17]

The 'removable confusions' in Hartlepool are either subtle and rarely recognised, or infamous and persistent. They arise when motives are obscure, and acceptable only for particular groups at particular times. They lose their meaning outside this context. They persist in constraints which distort the process of change and in contradictions in the legal and political context, which confuse the goals of change.

Plans, policies and methods have changed since the 1950s. Centralisation of government, comprehensive redevelopment and suburbanisation have lost their appeal. Despite this, certain goals and values are constant. In Hartlepool, people couch their needs and desires in terms of the functioning of material forms and the economics of physical change. On the whole, they do not see beauty in architectural space *per se*, but in the home and its associations – continuity, comfort, economy and, in particular, nostalgia.

No general consensus exists either for or against conservation. Views are not divided along party political lines, but the promotion of conservation is largely confined to the local Civic Society, the Member of Parliament, and various officers of the Council. National legislation and policy for conservation (to which Cleveland County Council subscribes), has had little tangible impact upon central Hartlepool. Conservation is devalued by 'cost–benefit' balance sheets, and is therefore not popular. Paradoxically, conservation is assumed by the ratepayers to be dependent upon the Council for political and financial support; yet it is rejected by members of the Council as not part of their responsibility. This contradiction has caused inaction in several cases and leads to indifference towards conservation as a general policy.

Highly significant and revered places in central Hartlepool are few, and most were created before 1939. Despite efforts to create a comprehensive town centre, features that are anathema to conservation – specialisation, standardisation and segregation – abound in Hartlepool. The streets have features in common: house symbols, advertisements, litter bins, street lights, poles, similar shop fascias, bus shelters, seats, bollards, railings and planters. These vary in age and material but most resemble the standardised designs depicted in the Design Council's Index of Street Furniture.[18] The Council supplies street furniture which is mass-produced by national suppliers, even though it has the discretion to buy local products. The streets in this study have become more specialised: wholesaling in Church Street, commerce in York Road, foodshops in the Middleton Grange Shopping Centre. The variety and vitality associated with traditional 'high streets' are missing, and few people appear to 'revel' in the streets. The Shopping Centre functions well commercially, but does not encourage lingering. York Road's architectural space is cared for, but there is too much traffic and not enough for people to do.

Pedestrian streets, adopted by the conservation movement, have been canvassed in Hartlepool.[19] Support for the idea, however, is muted. Most people support the principle but are anxious about its effects on commercial prosperity, the efficiency of the docks, and the consequences of re-routing traffic onto streets previously undisturbed by heavy traffic. In York Road, there is more support for the proposition of a buses-only route but the proviso about diverting traffic still holds. Indeed, the evidence indicates a broad acceptance of current planning ideology rather than adherence to an alternative ideology of conservation.

Not only are the values that underpin these views apathetic or even antipathetic towards conservation, the concept of conservation is itself misunderstood. I was told on one occasion that *'Hartlepool disappeared before planning started'*; a comment that implies that planning would have saved the town and is synonymous with conservation. That Hartlepool disappeared as the *effect* of planning is evident from its history: already the town's centre has been demolished once and moved twice, instigated by grand plans and concerted action. The changes that ensued were often contradictory. Lock, for instance, planned to relocate ancillary services and rehouse people on the outskirts of the town. By doing this, he suggested that open areas could be created to bring green space and fresh air into the 'smoke-ridden' town centre. But since almost half the population living in the area had to move, who was to be left to breathe the fresh air and what industry would remain to pollute it? What was the consequence of this solution on the 'strong community spirit' that Lock himself had praised?[20] The analogy that ran through the plan likened the town to a living organism, yet the 'town' was reified at the expense of the people living there, forgetting that the town would cease to exist unless people continue to live, grow and organise themselves there.

Planning based on two-dimensional space is inadequate for conserving places. Plans and maps are a reflection of existing or past situations. They cannot project the future without ignoring the process and the experience of change. In particular, the graphic artists' published impressions, derived from Lock's proposals, conveyed images that have proved since to be an unsound basis for evaluation by decision-makers or by the general public. If drawings idealise the anticipated results, then criticism is forestalled. Physical design is still seen as a means of social engineering. Visions of the future remain formulated in grand schemes for redevelopment, despite the lack of enthusiasm for the results of previous exercises. Perhaps statutory planning is no better than private enterprise in caring for streets and, more importantly, for people.

Little has been done to foster the care of streets. Private initiative attends to individual buildings, but money is not spent unnecessarily. Private property appears in practical basic colours. Brick facades are mostly left unpainted and unrendered, pipes are rarely painted out. Architectural detail is often ignored during redecoration. Fascias are either new and standardised or old and worn. In short, few exceptional efforts attract the eye, and personal responsibility for the streets is dormant except for the group of traders in York Road, whose care extended beyond the bounds of their own premises and into the street. The local authority has not conserved the streets I studied. The motifs are not grouped to construct an imagery of conservation, but rather indicate the local authority's priorities, in which economic factors are valued more highly than pedestrians' needs other than safety.

Contradictions exist between the national view of the architectural space in Hartlepool and the view of the Borough Council. The Council has not used its powers of compulsory purchase to protect the Wesley Chapel (a listed building) from neglect. In 1968, Christ Church was bought for a nominal £100 with a covenant against its demolition. Twelve years later a report on its structure and fabric is still pending. In the meantime it remains locked and unused, regardless of its potential value as a focal point and meeting place for secular purposes. The Council has made no use of the provisions of the 1968 Transport Act, which permits closure of streets to motor vehicles on grounds of amenity. The Council blames its tight budget for inactivity, but other boroughs do not show the same restraint.[21] The evidence suggests that it is the councillors' predisposition for or against conservation which determines the time lag.

Similar contradictions occur between the County and Borough Councils. Cleveland County Council is concerned about the town's dereliction and neglect. Hartlepool Borough Council is worried about the provision of new facilities where none currently exist. The Borough Council, for example, has applied to the Minister for the Environment for permission to demolish the Municipal Buildings, which are listed as historic buildings. This has been contested by the County Council, and the Minister's decision is in abeyance while the Borough Council undertakes a review of all listed buildings in Hartlepool.[22] The Borough Council has rejected an offer by a private developer to convert the Municipal Buildings into a hotel, in anticipation of the Highway Engineer's road plans, which demand their demolition.

These contradictions in policy, and the conflicts which ensue, contribute to the lethargy over conservation. More significant but less obvious constraints stem from economic myths. Capital may be the apparent cause of eyesores (see p. 150), the decay of the streets and the threat to places with social significance; but this is a mythical cause that acts as an excuse for designer, developer and councillor alike and allows them to follow the path of least resistance. As Scruton[23] describes, this path leads to the 'horizontal style' of building: to the transparent form that substitutes the life within for the 'life' of the facade itself and, I might add, in the case of street furniture, to the stereotyping of items and functions which supplants the imagination of vernacular design. It is a myth that economics rules the designer; abstract concepts do not take decisions. Scruton[24] maintains that modern 'horizontal style' does not fit streets, with no top or bottom, front or back to create a facade. Moreover, it is not even technically, materially or economically necessary since designs based on the 'principles of vertical organisation' are more appropriate for commerce.[25] The same arguments may be extended to street furniture. Despite the universal argument that funds cannot be spent on trivia, the decisive factor is not the availability of money but rather the attitudes of those who dispense the money. They decide how to care for streets, and they choose whether or not to stereotype forms and images. This disjunction between pragmatism and ideals is at the heart of this chapter: that places are not all that people want; that changes do not achieve what people imagine; that the streets are confused in their function and compromised by the agents of change; and that conservation is distorted in the political process between some people's ideals and other people's interests.

The absence of the imagery and ideology of conservation suggests that Hartlepool was not party to the 1960s revival of civic pride. However, there are signs of change. Now, an awareness of conservation underlies current action, even if there is no consensus for a concerted policy. Buildings and street furniture are left or adapted for new functions, whereas previously they were destroyed. This change marks some infiltration of the ideology of conservation in Hartlepool, although so far the impact is insignificant. To be effective, conservation requires an honest appraisal of problems and an integrated process of change to places. Instead, the agents of change proceed with set plans for physical change and their methods and motives are left largely unquestioned. This perpetuates the established ideology of planning and effectively discourages the adoption of a conservation ideology which is sensitive to the identity of places.

Relph has blamed the agents of change for 'the casual eradication of distinctive places and the making of standardised landscapes that results from an insensitivity to the significance of place.'[26] He suggests that the failure stems from the state of mind which motivates environmental change and that a remedy may be found through individual responsibility for place-making, whose 'order must be derived from significant experience and not from arbitrary abstractions and concepts'.[27] However, I have argued in this chapter that an understanding of individual experiences with place also requires an understanding of the social context and social values. Conservation of the built environment is an historically specific ideology of change, which represents social action based on shared ideals and values. The ideals of conservation are relatively stable tenets which infer notions of harmony and stability; standards of comfort rather than luxury; concepts of complexity and unity; ethics of respect and foresight; the morality of responsibility and equality; the balance of freedom and control. These are all properties of the temporality and identity intrinsic in places. Personal identity is conferred by 'place' and enhanced by the dialectic of past, present and future time, by the coexistence of 'transience and permanence'.[28] Hence, conservation of the built environment should represent the beliefs of the inhabitants and promote an understanding of the past and the present.

Raymond Williams' assertion that there is 'an amalgam of financial and political power which is pursuing different ends from those of the local community but which has its own and specific rationale'[29] is only part of the problem in Hartlepool. The interests of the local community have been lost in the reformation of the architectural space of the town's central streets. Although local people were party to the changes that were made, the fundamental problems in legislation and planning ideology were beyond their control. People understand the process of change but they are much less clear about what they think the changes were in the past or are now intended to achieve. They have lost the right to take individual action. Planning controls seem to have pre-empted almost all anarchic elements of life. I do not advocate licence for complete anarchy, but rather freedom to make statements of care for the street on an individual basis. This would be possible in a context where the goals of conservation are understood without statutory enforcement which standardises the appearance of public space. The ethos which enshrines concrete, durable changes should also embrace ephemeral but nonetheless creative action, reflecting with flexibility and variety the values of the community and reinforcing the identity of places.

Acknowledgements

The author acknowledges the help of Jacquelin Burgess, Suzanne Mackenzie, Charlotte Oliver and Richard Davidson in preparing this paper.

Notes

1 For example, European Environmental Bureau 1978. *One Europe, One Environment: A Manifesto*, 2nd edn, Brussels: E.E.B.
   See also Oliver, K. A. 1979. *Conservation Ideology and Environmental Change: a Study of the Civic Trust in England, 1977–78*. Unpublished M.A. dissertation, Department of Geography, University of Toronto.
2 The area referred to as 'Hartlepool' in this paper is the Borough of Hartlepool, which since 1974 has been part of the county of Cleveland. The comments of this paper, however, are directed towards that area formerly known as West Hartlepool.
3 Pevsner, N. 1953. *The Buildings of England: County Durham*, 158. London: Penguin.
4 The population of New Stranton in 1831 was 714. This grew to 11 736 by 1851 for what was then called West Hartlepool, and 63 000 by 1901. The 1980 population of the Borough of Hartlepool is approximately 100 000.
5 Cullen, G. 1961. *Townscape*. London: Architectural Press.
6 ibid.
7 *Mail* (Hartlepool) n.d., manuscript.
8 It is feasible that Cleveland County Council's influence over the streets' aesthetic may strengthen in the future.
9 Respectively Clifford Culpin and Partners as architects and Hillier, Parker, May and Rowden as property managers and development consultants.
10 Borough Engineer, Hartlepool Borough Council 1979. Personal communication.
11 Hartlepool Borough Council 1973. *Hartlepool Shopping Centre*. Unpublished manuscript.
12 Cement and Concrete Association n.d. Press release.
13 *Mail* (Hartlepool), 27 May 1970.

14 Hartlepool Borough Council n.d. *Official Guide*. Hartlepool: Borough of Hartlepool.
15 Hillier, Parker, May and Rowden 1979. *Local Authority Bulletin*. London: Hillier, Parker, May and Rowden.
16 *Mail* (Hartlepool), 12 September 1979.
17 Wheelwright, P., quoted in P. Hughes and G. Brecht 1979. *Vicious Circles and Infinity*, 75. London: Penguin.
18 Design Council 1976. *Street furniture from the design index*. London: Design Council.
19 For example, *Mail* (Hartlepool), 24 August 1979.
20 Lock, M. 1948. *The Hartlepools: A Survey and Plan*, 15, 80, 82. Hartlepool: Corporation.
21 Norwich, for example, had closed London Road to traffic as early as 1969.
22 Department of Environment 1979. Personal communication.
23 Scruton, R. 1980. Broadcast script, Radio 3, 20 September.
24 ibid.
25 ibid.
26 Relph, E. C. 1976. Preface. In *Place and Placelessness*. London: Pion.
27 ibid., 146.
28 Nairn, I. 1965. *The American Landscape: A Critical View*, 17. New York: Random House.
29 Williams, R. 1973. *The City and the Country*, 292. London: Chatto and Windus.

9 *The value of the local area*

SUSAN-ANN LEE

Introduction

This chapter considers the role and value of the local area to the individual citizen both in terms of his early psychological development and his adult social and political life, referring particularly to the nature and impact of current urban policies. Its main aim is to examine an assertion that lies at the root of many criticisms that have been directed against the areal basis of British urban policy: namely, that the concept of the 'local area' is no longer relevant to the urban resident and therefore no longer valued by him.

The chapter is divided into three main parts. The first looks at the relationship between place, especially local place, and the development of personal identity, considering some of the psychological aspects of an individual's sense of belonging to a place. The second part shows the continuing importance of the local area for the city resident's social life and its implications for urban policy and design. The third part examines the significance of the local area for urban political life, focusing on those situations in which the relationship between the individual citizen and government affects the citizen's ability to influence policy or design decisions about his valued local physical environment.

The role and value of the local area to the individual

The identification of places is a vital part of the early development of each individual. It is essential to our ability to orient ourselves and to move around in the environment, and hence to survival itself. As English and Mayfield[1] suggest: 'Space in a philosophical sense is empty. It requires bounding and identification by an individual, an interaction between self and environment, to be recognised.' Places have to be identified in space, wherein they become: 'goals or foci where we experience the meaningful events of our existence, but they are also points of departure from which we orient ourselves and take possession of the environment.'[2]

This quality of place as a focus and as a point of departure is important to the development of the child's knowledge of a 'system of places' in space, or a 'sense of place'.[3] Starting from his home base, the child explores outwards. He has both to learn to identify places and to distinguish between himself and the outer world.[4] Crucial to the latter is the child's active interaction with the immediate local social and physical environment.[5] Partly because of his size, but also because there are natural limitations to his self-induced mobility, the child only perceives a small-scale physical environment, which is itself a crucial part of the learning process. Indeed, it has been demonstrated by Mercer[6] that small-scale differences in the housing environment can significantly affect child-rearing practices and the child's ability to explore, which in turn affect his development.[7] Undoubtedly, the local environment of childhood is extremely significant for the development of both individual identity and a sense of place, but many designers have neglected these important relationships between people and place. Commenting upon this failing, the architect Sir Hugh Casson[8] noted that:

> Vaguely and quite rightly we know that a sense of place is the reinforcement of identity and thus, because the individual is thereby dignified, it helps indirectly to guarantee personal liberty. We know too that modern architecture has by and large failed to provide these qualities and we have turned our faces elsewhere.

The development of a strong emotional or symbolic attachment to a particular place – identification *with* place – is a separate process to that of the identification *of* place described above. Many criteria, both favourable and unfavourable, are used to identify places, but identification with place or 'a sense of belonging' to a place refers to that particular relationship between an individual and a place that has been termed 'topophilia': 'the affective bond between people and place or setting. Diffuse as concept, vivid and concrete as personal experience'.[9] Although identification of and identification with place are separate processes, they are often connected in early childhood. This is because the places first identified, such as 'home', have emotional significance and come to be highly valued, since they form the initial and stable framework for further environmental and social exploration and development.

The development of this sense of belonging to a place is imperfectly understood, although it is acknowledged to be important. For instance, Canter points out in his comprehensive review of studies of the psychology of place[10] that:

> There is growing awareness that one of the most significant properties of a place is its direct personal relevance to the person . . . in other words, the degree to which, if at all, the conceptualisation which a person holds of himself overlaps with the conceptual system he has of the place.

Nevertheless there are some writers who have studied this affective relationship, amongst whom are Clare Cooper[11] and Felizitas Lenz-Romeiss.[12]

Lenz-Romeiss describes in detail how an individual develops a symbolic attachment to place, relates this to the person's wider social life and draws out the implications for town planning. Clare Cooper has studied the house as symbol of the self and is also currently developing such techniques as the 'environmental autobiography', which elicits an individual's environmental values and their origins.[13]

There is some evidence that the *urban* aspects of an environment affect both identification of and identification with place. Ward[14] and Goodman[15] both vividly describe the impact (positive and negative) of cities on the growing child. In terms of the impact of the city on adults, there are writers who argue that the modern city has grown so large and is changing so fast that the individual is losing his identity, or fleeing to the suburbs to retrieve it. In contrast, there are those who argue that the full potential of the individual can only be realised in the rich diversity of modern city life. Alexander[16] and Sennett,[17] for example, argue that the flight to the suburbs is psychologically detrimental not only to those left behind, but also to those who flee. Sennett dramatically outlines the psychologically and socially repressive implications of 'assumed homogeneity' in the resulting residential neighbourhoods. Certainly it is true that the modern city dweller need not maintain diverse social relationships in order to survive, nor is he tied to the immediate physical locality for social contacts. It is also argued that reference groups (those on whom the individual models his behaviour) are becoming less 'local' and more 'national' in the urban environment.[18] From this one might conclude that while the immediate 'local' area has an important role in early social and psychological development and is therefore highly valued at that time, it is no longer important to the more general social life of the adult in the city. It is to this matter that we now turn.

The role and value of the local area in social life

The work of Melvin Webber[19] has perhaps done most to popularise the view that the increase in physical and social mobility in the 20th century would lead to the establishment of interest-based communities rather than place-based communities. Social life would revolve around those with whom one shares common interests and these people need not be in the immediate locality: indeed often they are not. This kind of social life, with 'community without propinquity',[20] has been called 'communality'.[21] Studies have shown that the mobile middle class is developing interest-based social lives, but research[22] also shows that this does not necessarily preclude social and political involvement in the local area as well. It is no accident that the pub, a key part of British social life, is known as 'the local'. More importantly, for the relatively immobile groups such as the old, the young and the poor – who comprise about half of the population – research shows that the local physical environment is of prime importance to their social life: it is their behaviour setting.[23] Many policy makers and designers seem to be unaware of this, coming as they do from the mobile middle class, yet it is here that their environmental decision making should be most sensitive, for these residents do not have the access to the options afforded by mobility. Lenz-Romeiss[24] describes how local social groups, whether mobile or immobile, come to be attached to place through what she calls 'place-related communication'. The 'local' dimension of this communication is crucial to the process, for 'communication is only conceivable in terms of social groups, and place-related communication, like symbolic attachment to place as a whole, is based on the existence of *local* groups.'

It seems that we are all involved to a greater or lesser extent in place-related communication in our local area, and thereby come to be attached to it. Research shows that this complex interrelationship between physical setting and social life was not fully appreciated by policy makers and designers in the neighbourhood and community planning of post-war Britain. Many studies have highlighted the significance of the local area to the social life of the urban working class, but perhaps none as vividly as those by Young and Willmott[25] and by Fried[26] when they were writing about those who were forced to move because of planned post-war redevelopment. Young and Willmott observed:

> [even when] planners have set themselves to create communities anew, as well as houses, they have still put their faith in buildings, sometimes speaking as though all that was necessary for neighbourliness was a neighbourhood unit, for community spirit a community centre . . . But there is simply more to community than that . . . If the authorities regard that [community] spirit as a social asset worth preserving, they will not uproot more people, but build the new houses around the social groups to which they already belong.

Fried describes the psychological reaction to relocation as 'grief'. He identifies the two main components of this reaction as, first, the fragmentation of the sense of 'spatial identity' and, secondly, the result of the dependence of the sense of 'group identity' on stable social networks:

> This view of an area as home and the significance of local people and local places are so profoundly at variance with typical middle class orientations that it is difficult to appreciate the intensity of meaning, the basic sense of identity involved in living in the particular area.

In the past it was just this kind of urban area for which major change was planned. To some extent, the bitter lessons learned from relocating their residents have contributed to the recent modifications in national and local policy, changing from large-scale urban redevelopment to smaller-scale rehabilitation of existing physical environments under such policies as Housing Action Areas and General Improvement Areas. This shift in policy is laudable because decanting is limited and public participation has become part of the process, but it can be seen that the social and environmental changes are still considerable at the local scale, and in many cases the residents lose their sense of belonging to that place. This even applies to areas where traditional urban working class communities no longer exist and the residents are less fundamentally tied to the locality.

We know surprisingly little about the informal social lives of urban residents and even less about how they build up social networks over time in a particular place. This is partly because the process itself changes if it is studied by an outsider. It is ironic that where policy makers and designers have to design completely new environments – for which an understanding of the relationships between people and place is particularly crucial – there is little feedback of information from previous new settlements to guide them. Certainly the 'natural laboratory' of British New Towns has hardly been tapped for this kind of study. The research that has been carried out indicates that the relationship between physical setting and social life is complex, but that it is important in the lives of residents. In addition, it has been shown that even the more mobile inhabitants value their local areas in both psychological and social terms. We will now turn to the value of the local area to the individual in political terms.

The role and value of the local area in political life

In contemporary Britain, the gaps between the government and the governed, between the management and the worker, and between the planner and the planned are widening. At the local level, the gap between the government and the governed has been widened further by the reorganisation of local government. This has meant that councillors make representations to a smaller number of authorities, which are larger than before reorganisation. Hindess[27] has shown that local 'government', which implies citizens' democratic involvement and debate over priorities, is increasingly seen by residents as local 'authority', implying the efficient supply of basic services.

Central government has adopted an 'action-research' approach to the development of urban policy in Britain over the last 20 years. Many research initiatives and subsequent national policies have been area-based: that is, not only is action to be located within an identified urban area, but action is also geared specifically to deal with the problems of that one area. The government sees the combination of action-research initiatives and area-based policies both as a promising approach to the solution of urban social and environmental problems *and* as a potentially invaluable mechanism for increasing the responsiveness of local authorities to the needs of their areas, by bridging the gap between the citizen and his local government. Area-based policies have been developed separately in housing, social services, planning and education, to tackle specific problems in the urban areas chosen. However, as a result of the findings of action-research projects including the Urban Guideline Studies[28] and the Inner Area Studies,[29] a more comprehensive, but still area-based, strategy has been put forward recently for urban areas. This strategy may be witnessed in the Inner City proposals,[30] Area Management proposals,[31] and Comprehensive Community Programmes.[32]

Although writers such as Hambleton[33] view area management as a mechanism to make local authorities more responsive to local needs, and as a way to improve communication between the individual citizen and his local government, others have criticised the area approach to urban policy on several grounds. The most fundamental criticism is made, *inter alia*, by Norris[34] and by Harvey,[35] who show that the implicit assumption of this approach – that urban problems have a recognisable spatial distribution with causal links – is an 'ecological fallacy'. Certainly there is little evidence to support this assumption. It is agreed that an area approach may alleviate some of the symptoms of urban problems in the chosen areas, but that it is unlikely to tackle the causes of these problems, which are not 'local', but more fundamentally rooted in the structure of society.[36] Yet in spite of criticism, it is likely that the government will continue to pursue area-based policies to tackle urban deprivation in specific, selected areas. As the political and planning framework for all other parts of the city also operates on an areal basis, it is pertinent to look now at how policy makers and designers define and value areas within the city as the basis of their decision making.

Political decision makers have thought traditionally in terms of the spatial framework of the ward, constituency, district and county. It is essential that government is locally democratic in those aspects that have greatest meaning for people's lives, but there is little appreciation by political decision makers that their own view of the area, its value and its needs may not coincide with those of the residents they represent.[37] Policy makers within central and local government have widely varying concepts of the local area and its value. Within a single local authority that operates several services on an areal basis, it is common to find that the areas of the city defined by the housing department for its services would differ from, say, those of the social services department, or those of the education department. Although it is unlikely that there is an 'ideal' areal division, a more corporate approach could eliminate the unnecessary negative effects of these multiple divisions. Research also shows that the policy maker's definition of the most needy areas of the city, for example, may be made on a political rather than a technical basis (i.e. one that is objectively measured by social research).[38] This has important implications for the citizens' perceptions and their involvement, both within and outside the urban areas so chosen.

Designers, planners and architects perceive areas within the city in the 'objective' terms of their specialism, both in terms of their extent (e.g. areas designated for 'local plans' and 'structure plans') and in terms of their content (e.g. housing, industry and routes). The study by Sewell and Little[39] shows that there are important differences in environmental perception between the professions, both in terms of problems and solutions, as well as between professionals and laymen. To some degree, an awareness of these differences lies behind the growing movement for environmental education[40] in Great Britain, which is not only enabling residents to have a better understanding of designers' differing views and values of their area, but also elicits residents' views of their area in ways which can inform both policy and design.

Despite the many studies that identify residents' perceptions of the extent of their neighbourhood,[41] this type of boundary definition has rarely been used as the areal basis for decision making. This is partly because methodological and theoretical confusion in these studies questions the validity of the areas so defined,[42] and partly because the general findings of these studies suggest that perceptions of local areas within cities are highly individual. In content, value or extent they are rarely coincident with the political wards, nor with designers' 'planned neighbourhoods', nor even with those as defined by their immediate neighbours. This poses a dilemma for policy makers in area definition for urban policy. It would seem that the most pertinent definition of the area would be that of the residents, but in many cases they are unlikely to agree as a group. Nevertheless, the fact that people do perceive 'local' areas as such is important. As Buttimer and McDonald[43] point out, in relation to the Community Attitude Survey, it 'may in the long run prove to have been of paramount importance in drawing attention to the potential of such studies for those engaged in planning at all levels of the system.'

The implementation of General Improvement Areas (hereafter GIAs) perhaps has come closest so far to a working integration of the attitudes of political decision makers, designers and residents in the context of a small urban area. GIAs particularly concern the individual citizen, his home and its local and valued physical environment which is recognised in the statutory responsibility under the 1968 Housing Act to include public participation. The definition of the extent and content of the potential GIA has usually been made by the local authority policy makers using available sources of physical and social data, but:

> two [other] authorities, in selecting their pilot areas took account of residents' views on what constituted their neighbourhood: although the choice of a large area – over 600 houses in one case and over 2,000 in the other – meant that implementation would have to be phased, it was felt that the boundaries of the area represented the boundaries of the community to which residents felt they belonged.[44]

Once the areal extent of the GIA has been defined, the subsequent planning may take one of three forms,[45] which vary in terms of the degree to which

local residents' views and values will shape the plan. In the first, a comprehensive plan is prepared by professional policy makers and designers,
presented to the residents at either an exhibition or a public meeting, or both,
and the details are worked out later after discussions with residents and
surveys. In the second form, the various possibilities are discussed at a public
meeting and then a specific plan is put forward by the Council. The third
form is more genuinely participatory. The area's deficiencies are discussed at
a public meeting, after which a residents' committee is formed to prepare a
plan in close co-operation with the Council's policy makers and designers.
This method was adopted by the London Borough of Hackney who consider
'that it represents a successful experiment in public participation.'[46] It
probably enables residents to have a greater influence in the decision making
for their local area than in any other situation within the context of British
political and planning systems but, of the small number of GIAs so far
declared, relatively few have adopted this approach.

Conclusion

As has been shown, although the local area is valued by the urban resident
psychologically and socially, this relationship between people and place is
imperfectly understood, and the mechanisms enabling residents to communicate these values to policy makers and designers are weak. This is slowly
changing as public participation measures are being adopted and their
inadequacies are being reviewed.[47] A fundamental problem remains,
however, in that participation measures are nearly always initiated by the
authority rather than by the citizen. The latter remains on the bottom rungs of
Arnstein's[48] 'ladder of participation', being only allowed a 'token' degree of
influence rather than delegated power or citizen control. There are few
occasions in Britain when residents can make their own policy and design
decisions for their local area. The situation is also changing slowly through
the growing environmental awareness of the public. This has arisen for two
reasons. In part it reflects the sheer scale of environmental change in Britain
over the last 30 years and the largely negative reactions that people have had
towards it; but it also is due to the spread of environmental education, even
though much more needs to be done before environmental education in
schools can produce more informed citizens, able to participate in environmental decision making, and more enlightened designers and policy makers.

Research into the relationships between people and place has been tentative
but its output has been steadily growing. One can see now that 'mobility will
never destroy the importance of locality'[49] and that, as Spencer[50] says, 'in the
last resort, it is not so much the tangible physical elements of places that are
important, as the values that people invest in them.' We have a long way to go
in our understanding of how people come to value places, and how physical

changes or relocation affect this, but at least it has come to be acknowledged as an important issue to which a start has been made.

Notes

1 English, P. and R. Mayfield (eds) 1972. *Man, Space and Environment*, 214. London: Oxford University Press.

2 Norberg-Schulz, C. 1971. *Existence, Space and Architecture*, 19. London: Studio Vista.

3 For the purpose of this paper 'sense of place' refers to an individual's mental structuring of places which enables the identification of a place within the system.

4 Piaget, J. and B. Inhelder 1956. *The Child's Conception of Space*. London: Routledge and Kegan Paul.

5 Goffman too stresses the importance of familiar environments to the adult's subsequent presentation of self, see Goffman, E. 1959. *Presentation of Self in Everyday Life*. New York: Doubleday.

6 Mercer, C. 1975. *Living in Cities: Psychology and the Urban Environment*. London: Penguin.

7 Newson, J. and E. Newson 1965. *Patterns of Infant Care in an Urban Community*. London: Penguin.
   Newson, J. and E. Newson 1970. *Four Years Old in an Urban Community*. London: Penguin.

8 Casson, H. 1973. Questions on competence. *R. Inst. Br. Arch. J.* **80,** 166.

9 Tuan Y.-F. 1974. *Topophilia: a study of environmental perception, attitudes and values,* 4. Englewood Cliffs, NJ: Prentice-Hall.
   See also Lynch, K. 1960. *The Image of the City,* 123–39. Cambridge, Mass.: MIT Press.

10 Canter, D. 1977. *The Psychology of Place,* 122–3. London: Architectural Press.

11 Cooper, C. 1974. The house as symbol of the self. In *Designing for Human Behaviour: Architecture and the Behavioural Sciences,* J. Lang, C. Burnette, W. Moleski and D. Vachon (eds), 130–46. Stroudsburg, Pa: Dowden, Hutchinson & Ross.

12 Lenz-Romeiss, F. 1973. *The City: new town or home town?* London: Pall Mall.

13 Done with a view to deepening understanding of how these values affect individuals, whether they are designers, researchers, or residents. See also Goodey, this volume.

14 Ward, C. 1978. *The Child in the City*. London: Architectural Press.

15 Goodman, P. 1958. *Growing Up Absurd*. New York: Random House.

16 Alexander, C. 1967. The city as a mechanism for sustaining human contact. In *Environment for Man: the next fifty years*. W. Ewald (ed.), 60–102, 292–6. Bloomington: Indiana University Press.

17 Sennett, R. 1971. *The Uses of Disorder: Personal Identity and City Life*. London: Allen Lane.

18 Frankenberg, R. 1966. *Communities in Britain: social life in town and country*. London: Penguin.

19 Webber, M. 1964. Culture, territoriality and the elastic mile. *Papers and Proceedings of the Regional Science Association* **13,** 59–69.

20 Webber, M. 1963. Order in diversity: community without propinquity. In *Cities and Space*, L. Wingo (ed.), 23–56. Baltimore: Resources for the Future.

21 McClenahan, B. 1945. The communality: the urban substitute for the traditional community. *Sociol. Social Res. J.* **30,** 264–74.

22 Litwak, E. 1961. Voluntary association and neighbourhood cohesion. *Am. Sociol. Rev.* **26,** 258–71.
Doherty, J. 1969. *Developments in Behavioural Geography.* Discussion Paper 35, Graduate School of Geography, London School of Economics.
Darke, J. and R. Darke 1969. *Physical and Social Factors in Neighbour Relations.* Working Paper 41, Centre for Environmental Studies, London.
Carey, L. and R. Mapes 1972. *The Sociology of Planning: a study of social activities on new housing estates.* London: Batsford.
Lenz-Romeiss, F. op. cit.
23 Barker, R. 1978. *Habitats, Environments and Human Behaviour.* San Francisco: Jossey-Bass.
Also Goffman, E. op. cit.
24 Lenz-Romeiss, F. op. cit., 45.
25 Young, M. and P. Willmott 1957. *Family and Kinship in East London,* 166. London: Routledge and Kegan Paul.
26 Fried, M. 1963. Grieving for a lost home: psychological costs of relocation. In *The Urban Condition,* L. Duhl (ed.), 362. New York: Basic Books.
27 Hindess, B. 1971. *The Decline of Working Class Politics.* St Albans: MacGibbon and Kee.
28 Department of the Environment 1973. *Making Towns Better: reports on Sunderland, Rotherham and Oldham.* London: HMSO.
29 Department of the Environment 1977. *Inner Area Studies: Liverpool, Birmingham and Lambeth:* summaries of consultants' final reports, London: HMSO.
30 Department of the Environment 1979. *Policy for the Inner Cities.* Cmnd 6845. London: HMSO.
31 Department of the Environment 1974. *Area Management.* Paper circulated by the Department of the Environment, September.
32 Home Office 1974. Statement by Home Secretary Roy Jenkins on Comprehensive Community Programmes, House of Commons, July.
33 Hambleton, R. 1978. *Policy Planning and Local Government.* London: Hutchinson.
34 Norris, G. 1979. Defining urban deprivation. In *Urban Deprivation and the Inner City,* C. Jones (ed.), 17–31. London: Croom Helm.
35 Harvey, D. 1973. *Social Justice and the City.* London: Edward Arnold.
36 The National Community Development Project 1973. *Inter Project Report.* Report to the Home Secretary, The National Community Development Project, London.
37 See Uzzell, D., this volume.
38 Duncan, S. 1974. Cosmetic planning or social engineering? Improvement grants and improvement areas in Huddersfield. *Area* **6,** 259–71.
39 Sewell, W. and B. Little, 1973. Specialists, laymen and the process of environmental appraisal. *Regional Studies* **7,** 161–71.
40 See for example, issues of *The Bulletin of Environmental Education.* Ward, C. and A. Fyson. 1969. *Streetwork: the exploding school.* London: Routledge and Kegan Paul. Goodey, B. 1975. *Urban walks and town trails: origins, principles and sources.* Research Memorandum 40, Centre for Urban and Regional Studies, University of Birmingham.
41 Spencer, D. 1973. *An Evaluation of Cognitive Mapping in Neighbourhood Perception.* Research Memorandum 23, Centre for Urban and Regional Studies, University of Birmingham.
42 ibid.
Also see Lee, S.-A. 1975. Cognitive mapping research. In *Responding to Social Change,* B. Honikman (ed.), 172–88. Stroudsburg Pa: Dowden, Hutchinson & Ross.
43 Buttimer, A. and S. McDonald 1974. Residential Areas: planning perceptions and

preferences. In *Studies in Social Science and Planning,* J. Forbes (ed.), 169. Edinburgh: Scottish Academic Press.

44 Department of the Environment, Welsh Office 1973. *Public Participation in General Improvement Areas.* Area Improvement Note 8. London: HMSO.

45 ibid.

46 ibid.

47 See, for example, the Linked Research Project into Public Participation in *Structure Planning Research Papers* 1–14. University of Sheffield.

48 Arnstein, S. 1969. A ladder of citizen participation in the USA. *Am. Inst. Planners* **35,** 216–24.

49 Pahl, R. 1970. *Whose City? and other Essays on Sociology and Planning,* 121. London: Longman.

50 Spencer, P. 1971. Towards a measure of social investment in communities. *Arch. Res. and Teaching* **1,** 32–8.

10 *Valued environments and the planning process: community consciousness and urban structure*

DEREK R. HALL

In examining valued environments in the context of an English city, this chapter will suggest that an environment can be considered 'valued' if its residents can show recognition of and empathy with their local social and physical environment, a milieu that in turn serves to buttress their own preferred life-styles. In this sense, the word 'value' acquires a mirror-like quality, in which residents recognise attributes in their local environment which are themselves reflections of the residents' own scale of preferences. What is the nature and significance of such 'value' and, indeed, is there merit in trying to articulate it? Further, if such 'valued' environments are worth protecting and enhancing – and by definition their residents must think so – then how can this be done? Is the present system of political representation in urban areas adequate to meet such a task, and able to voice sufficiently the needs and aspirations of what may well be a relatively large number of small and disparate valued environments? Could micro-scale urban neighbourhood councils, elected on a street-by-street basis perform such a task? These are the questions to which this chapter is addressed.

The subjective reality of valued environments

The evaluation of any environment is perforce subjective, and attempts to compare different evaluations of the same environment encounter considerable difficulties, whether done quantitatively or qualitatively. In this essay, the nature and significance of valued local environments are assessed by examining the attachment of residents to their home area or their 'local community consciousness'. Perhaps the clearest exposition of the major components of such a concept is supplied by Burnett:[1]

> sharing common interaction and involvement (activities) and having a
> sense of common identity and interests (attitudes) are significant

ingredients. Interaction refers to the social network of face to face relationships amongst friends, relations, acquaintances and neighbours which may occur in a local area with varying degrees of frequency and intensity. Involvement implies the active participation in institutions and organisations (and assumes their existence). Identity consists of those feelings of belonging and attachment for places that surveys have demonstrated exist to varying degrees. Finally, interests refer to common or at least compatible values and needs (often expressed as grievances) which people feel they share on a number of issues which confront them. When a group is integrated and its activities and attitudes tend to converge in the same local area, then this constitutes a local community. Such a characterisation may not be entirely satis-factory, but it does contain both a social and spatial dimension, and includes both the facts about peoples' lives as well as self awareness of community membership that many deem necessary in a valid definition of the term.

To investigate the role and importance of such consciousness and, in particular, to discern what such attachment tells us about the valued environ-ments themselves, the author undertook a large-scale survey of residents in the city of Portsmouth on the south coast of England (population 200 000). A questionnaire containing 29 questions was administered to a sample of almost 600 respondents. At its core was the question:

Is there an area around here in which you live, that you can call a 'community area'?

If an affirmative answer was received, a further three related questions followed. Residents were asked to mark this area on an accompanying map – a method known as the 'structured graphic technique'[2] of cartographically representing local area attachment. The maps which were used were duplicated copies of the Portsmouth town map[3] segmented into eight over-lapping sheets. Of course, even assuming that residents hold perceptions of their local area, they may be unfamiliar with using maps, may be diffident when asked to show their drawing skills, and indeed may hold perceptions of their environment which are not amenable to cartographic representation, such as smells, feelings and sensations. Moreover, when using such printed base maps, interpretation depends not only on area knowledge, but also upon recognition of the symbolic language being employed. Such symbols, for example, can act as 'environmental cues'[4] which bias responses.

In attempting to overcome these problems, residents were also asked to pinpoint the reasons for their attachment to the area by choosing terms with which to reveal the dimensions of their community area consciousness. These include such characteristics as 'live here', 'friends', 'relatives', 'certain types of houses', 'well known by name', and 'easily recognisable boundaries'. Finally, residents were asked if there was a name they gave to their area. Such names were subsequently related to various published 'cues', such as the recurrence of names on maps and in bus timetables.

Of the sample, 43 per cent were able to articulate a local community area consciousness. The size of the areas people recognised generally supported previous findings, which indicated that in towns and cities the area to which people feel attached often encompasses a number of streets with a population of up to 10000.[5] A large proportion of community areas (71 per cent) were larger than a single street but smaller than the equivalent of a ward, with 92.5 per cent of all areas being smaller than a ward.

From these results two different types of cognitive maps were produced: generalisations of areas articulated by at least ten per cent of the area sample interviewed (Fig. 10.1), and a contour surface of community area recognition (Fig. 10.2) produced by superimposing all of the cognitive maps to reveal common boundaries and to emphasise those areas most readily recognised by their residents.

What do such research findings tell us about people's attachment to valued local environments, and indeed what do they tell us about those environments themselves? It has been suggested[6] that 'boundaries' within the urban built environment, such as main roads, railways and other major changes in land use, are often employed to define local urban areas by those (e.g. planners) who are not acquainted with an environment through residence. More subtle features – a bend in a road or a specific building – may be used by local residents to locate the limits of their local area, thus revealing the importance, to long-standing residents, of micro-features within the urban environment. From the survey, some of the areal boundaries that were recognised were unambiguously defined on the map by their residents *en masse*, apparently reflecting widely held and easily recognisable boundaries to a valued environment. Other areas, however, were defined far less precisely, producing a feature which can be referred to as a 'frontier', representing gentle gradients on the contour map (Fig. 10.2) rather than sharp boundaries between adjacent community areas.

Thus characterised by an absence of perceived boundary markers, 'frontiers' reveal at least four degrees of relationship between spatial behaviour and the urban environment. First, they may exist where open access from adjacent community areas is freely available (such as local parks); secondly, where public access is implicitly available from both local and non-local area residents (such as cemeteries); thirdly, where access is essentially from non-local residents and is for a specific purpose (such as major hospitals); or, lastly, where large-scale demolition and dereliction have temporarily rendered an area devoid of any local use such that public access is irrelevant, except for use as a through route to other destinations. As such, locally perceived 'frontiers' represent situations whereby areas are 'shared' by

Figure 10.1 Community areas identified by more than 10 per cent of sample.

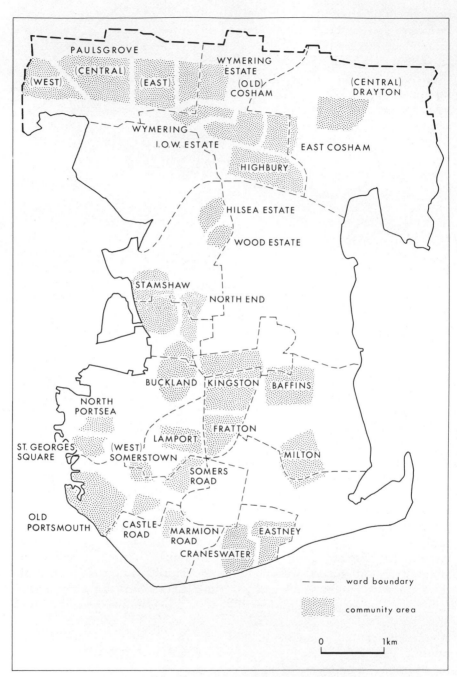

Figure 10.2 Contour surface of community areas.

residents of different territorial allegiances, and cannot be dominated by and be looked upon as territory belonging to one adjacent residential area. These areas, therefore, are not 'valued' in the same manner as the strongly articulated local community 'territories'.

Such perceptions stem from actual resident behaviour – interaction and involvement – within the local environment. Even distinctive and well bounded urban areas do not by themselves necessarily induce local interaction and involvement, as revealed by early neighbourhood planning in British new towns, with its heavy overtones of architectural determinism.[7] Indeed, from the author's survey, three types of ecological area failed to reveal a significant community consciousness. First, in several high status, low density areas which were characterised by high personal, social and physical mobility, the concept of 'community' appeared to have little meaning for residents within the locally defined area. This exemplifies the concept of a 'community of limited liability',[8] wherein residents only take a limited interest in their area of residence, due either to the fact that the focus of their activities lies elsewhere (e.g. at the place of work, recreation area or shopping centre) or because they only live in one place for a relatively limited time and do not develop attachment to it, or both. Secondly, high density areas of subdivided dwellings which were experiencing a high turnover of population also failed to reveal local attachment. Lastly, medium density areas of lower middle income groups, comprising terraced houses, built in four- or six-to-a-block style between the wars, also revealed little community consciousness, by virtue of both the lack of environmental character and any focus for local interaction and involvement, such as small shops and pubs.

For varying social, psychological, economic and physical reasons, these areas were not seen as valued environments by their residents. In contrast, the survey revealed strong local community consciousness in four other types of ecological area. First, strong attachment was found in middle class urban 'villages', where physical distinctiveness lends character to a valued status area, producing residential fashionability for a relatively homogeneous social group within a well defined area. Community consciousness was also found to be strong in high density, low status areas of closely packed Victorian terraces that have been unaffected by redevelopment, where time and social similarity has stimulated valued interaction. Thirdly, in certain municipal estates, the provision of communal social facilities such as a community centre was seen to be important in moulding a valued environment out of areas of relatively uniform housing and similar social groups. Finally, strong attachment was to be seen in medium density areas which had well defined physical boundaries and a high degree of internal cohesion, resulting from, for example, the presence of an active, undogmatic parish church. At the aggregate level, it therefore appeared that there was little discernible relationship between an area's socio-economic status and the degree to which its inhabitants articulated a community area consciousness. Various factors were important in the wide range of differing valued environments.

However, it can be claimed that social groups possess different and often conflicting interests (the neo–Marxian conflict approach to social organisation exemplifies such a philosophy). It has been suggested[9] that such analyses of differing environmental experiences should initially consider the actual language employed to express those experiences. In this context, Bernstein's work on class differences in language[10] has been applied by Goodchild in his study of variations in people's environmental images.[11] He pointed to Bernstein's distinction between 'public' language – short, grammatically simple sentences of poor syntax – used by most people, and 'formal' language – accurately formalised grammatically and syntactically – used only by managerial and professional status groups. The difference between them is primarily a function of differing levels of conceptual ability. It can be argued that those who are restricted to a 'public' code are less likely to conceptualise aesthetic aspects of the environment, being preoccupied with the basic needs of day-to-day life. This is certainly one factor which could support the contention that higher status groups will more readily reveal community area attachment, if only because they can more readily understand and articulate the concepts involved. Although no strong statistical relationship was found, the survey results showed that there was a general trend for lower status area residents to emphasise a greater proportion of factors connected with social interaction when articulating the characteristics which made up their local community area. By contrast, the residents of higher status areas tended to select physical (and implicitly aesthetic) attributes.

Looking more closely at some of the components of a valued local environment, it might appear that primary schools and shops would play an important role in stimulating local interest and interaction, since these are facilities used in common by local families. This role is often complicated, however, by the fact that the catchment areas of such local community components need not coincide with residents' perceived community areas. Functional catchment areas have traditionally been employed as a means of attempting to delimit neighbourhoods or communities. Indeed, post-war town planning has often incorporated the neighbourhood concept, constructing residential areas around a focus of such service activities. Yet, in reality, the catchment areas for these functions only rarely coincide, and each in turn tends to fluctuate over time. Indeed catchment areas for primary schools are officially vaguely demarcated. Whole streets are regarded frequently as 'frontier' areas, not static in themselves, due to the degree of choice of school under the 1944 Education Act. Out of necessity, they fluctuate according to the availability of premises and to the size and character of the local child population. Old premises tend to be found on or near main roads, whereas new ones are sited away from them, thus creating different catchment patterns which are further complicated by the existence of denominational schools. It can be argued that rather than alluding to areas of common community, primary school catchments may therefore cut across them, their spheres of influence not generally being related to that of other services. Moreover, in low status areas, this division between school and community activity may be further emphasised by the social dissimilarity of parents and teachers, who will not only live in different areas, but will also fail to share common values and experience.

Similarly, distributions of retail centres and service areas would also not appear to be particularly useful in helping to delimit and buttress areas of community consciousness. While the largest and most prestigious retailing outlets may be located in relation to areas of highest potential purchasing power, and certain low order retail provision – e.g. pawnbrokers or second hand clothes shops – tend to be associated with specific socio–economic characteristics, the overall nature of retail structures and hinterlands generally cuts across social differentiation. Indeed, except for the most specialised of provisions, an element of social balance within an outlet's catchment area would appear to be desirable.

It is perhaps not surprising that, in terms of their perceived importance in contributing to a local community area consciousness, the questionnaire survey revealed a relatively low recognition both for shops (mentioned by 45 per cent of those recognizing a local community area) and schools (22 per cent). Such facilities also varied considerably in their perceived relevance from place to place. For example, in two private housing areas residents did not mention shops or schools at all. Nevertheless, considerable importance was attached to shops in areas containing distinctive streets of antique shops and ancillary services, as well as in districts possessing shopping streets as area boundaries, and in the major shopping centres.

Representation of valued environments in the planning process

Valued environments are not static. Whatever method is used to define them, valued environments may physically expand, contract, deteriorate or be enhanced. Also, by definition, residents of such areas will wish to protect, enhance and give voice to the needs and aspirations of their valued environment. At present, this can be performed in three general ways. Central and local government services may be organised in such a way as to correspond in their areas of operation with a local community, with perhaps an area-based team working in such a way that this areal designation at least buttresses and enhances the recognition of a valued environment. Secondly, needs and aspirations can be voiced through locally elected representatives. Thirdly, such needs and aspirations may be articulated through variously constituted voluntary organisations.

The areas that local authorities and other statutory bodies demarcate for the organisation of their services rarely coincide in size or shape. They are usually constrained by such considerations as urban morphology, population characteristics, and availability of finance, personnel and office space – all factors themselves partly constrained by the underlying pattern of land costs. In 'action' areas established by central and local government to conserve, improve or regenerate distinctive and valued environments, residents' activities and interests are usually greatly affected. Local organisations seeking to participate in decisions affecting the area may be encouraged by the local authority, although the latter remains in the position of controlling and allocating resources, and of mediating on decisions over conflicting views and needs. Indeed, the fact that such areas need this form of intervention is often the result of non-local forces.

In attempting to improve the environmental quality of areas perceived to be of a particular standard, local authority General Improvement Areas, established under provisions of the Housing Act of 1969, reflect what has been termed the 'inverse care law'.[12] This is the principle whereby resources are often channelled into those areas which can be redeemed within the constraints of resource availability, rather than those areas which may be in greatest need (at which the provisions of the Housing Act of 1974 were expressly aimed). Yet, rather than causing more disruption to residents through comprehensive redevelopment, especially in areas of high owner occupation, General Improvement Areas have been seen by some observers as the panacea for economically cheaper, politically safer, and socially less harmful urban residential improvements. Despite this, sharp divisions of attitude have often revealed themselves within local authorities. In Portsmouth, for example, divisions of opinion existed not only between and within political parties represented in the council chamber, but also within and between competing local authority departments themselves. One may suggest a whole range of reasons for residents feeling that the needs and aspirations of their local area are not adequately represented in formal political processes. For example, the perception of inadequate representation from city councillors may derive from a variety of factors. These would include the councillor's personality and reputation as a ward councillor; his allegiance to a different political party from that of the resident, or belonging to a minority party with little say in council resolutions; social differences between councillor and resident; a councillor's non-residence within the neighbourhood, ward or city; and the effect of a ward boundary running through an area, and politically dividing its shared characteristics in such a way that councillors in the adjacent wards (which may be of opposing party affiliation) are perceived to be unable to view or represent the problems of the area *in toto*.

Such 'democratic' shortcomings may be further exacerbated in any valued environment by shortcomings in voluntary associations: their absence, perceived lack of political weight or unsuitability, or their apparent undemocratic nature. While it was previously shown that valued environments covered a wide spectrum of social groups and ecological areas, the social and spatial dimensions of political representation of these areas in Portsmouth at least was shown to be far from equitable.

In 1973, Portsmouth undertook the first elections for city representatives to the new Hampshire County Council, and the candidates at this election largely constituted the elite of the city's politicians. Some had already been elected to Portsmouth District Council, while others had been waiting in the wings for this larger political arena. To gain an idea of the social composition of Portsmouth's political representatives, an analysis of all candidates' stated employment was undertaken.[13] This revealed that, in terms of socio-economic status, the representatives were far from typical of the population as a whole. Managerial and professional groups were far in excess of their relative size within the population of Portsmouth (48.5 per cent compared to 11.5 per cent of the total population), while blue collar workers were vastly under-represented (27.2 per cent compared to 73.8 per cent). To complement these findings, an analysis of the residential distribution of councillors and aldermen was undertaken for periods before and after local government reorganisation to examine their spatial representativeness, respectively for the years 1971–2 and 1974–5. The numbers of city councillors remained the same for both analyses, but the pre-reorganisation structure of an alderman for each ward together with five honorary aldermen was ended by the Local Government Act of 1972. While the average distance between a councillor's place of residence and the closest boundary of his ward (where not actually residing in it) remained fairly constant for the two time periods, the residential disparities between those groups least represented and their elected representatives increased quite dramatically – from 1.3 kilometres to 1.8 kilometres for city councillors and to 1.5 kilometres for county councillors.

Three major spatial implications arise from this (Figs 10.3 and 10.4). First, a group of inner city wards (usually Labour represented and residentially less desirable) failed to have a single member of Portsmouth's elected representatives as an area resident. Secondly, and conversely, certain outer wards (Conservative represented and residentially desirable) contained an excess of council members living within them. The ward of St Simon in Southsea was an especially good example of spatial elitism, containing 11 Conservatives in 1971–2 and eight in 1974–5, along with the city's Town Clerk/Chief Executive. Thirdly, a large number of council members lived at a considerable distance from their constituencies, a distance enhanced by the contrast in environmental conditions between their home environment and the ward that they represented. For 1974–5, for example, the two Labour councillors for the inner city ward of Portsea, adjacent to the Dockyard, lived in Paulsgrove on the mainland, while the Conservative representative resided in Southsea.

Although it is not claimed that spatial distance per se is a major barrier in

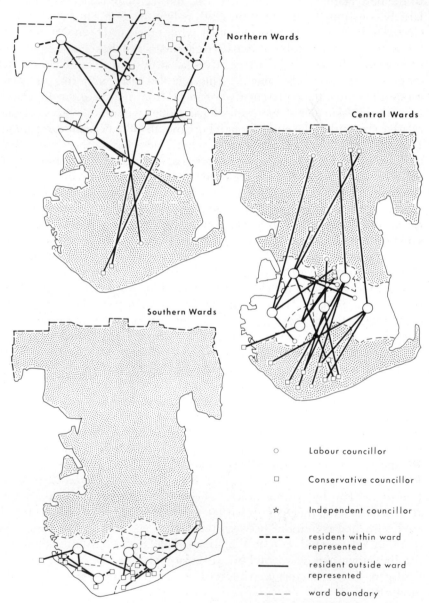

Northern Wards

Central Wards

Southern Wards

○ Labour councillor

□ Conservative councillor

☆ Independent councillor

‒ ‒ ‒ ‒ resident within ward
represented

━━━━ resident outside ward
represented

‒ ‒ ‒ ‒ ward boundary

Figure 10.3 Residential distribution of councillors and wards represented, 1971–2.

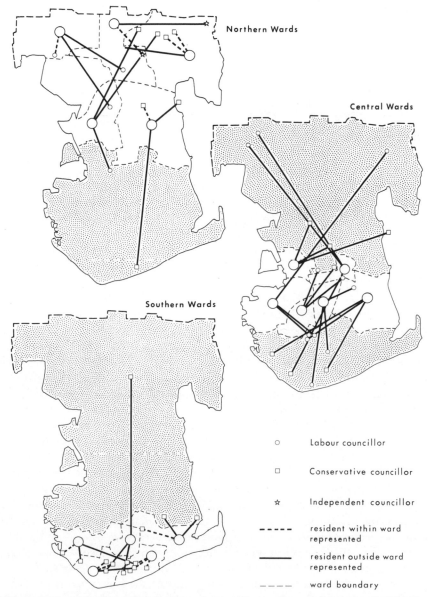

Figure 10.4 Residential distribution of councillors and wards represented, 1974–5.

Northern Wards

Central Wards

Southern Wards

○　　　Labour councillor

□　　　Conservative councillor

☆　　　Independent councillor

- - - - -　　resident within ward
　　　　　represented

──────　　resident outside ward
　　　　　represented

- - - -　　ward boundary

socio-political representation, these imbalances in the local democratic structure do imply that people living in low status contexts may not have their needs and aspirations sufficiently articulated in political terms, simply because political representatives lack residential experience of these areas. Such a situation is susceptible to the argument that neighbourhood councils, composed of people living in the local area, might be better able to articulate at first hand the needs and expectations of that area.

Neighbourhood councils and valued environments

At the time of writing, neighbourhood councils in England have been neither accepted by statute nor widely adopted. Nevertheless, a developing body of empirical observation can be drawn upon, given statutory provision for community councils in Scotland and Wales, the emergence of variously comprised non-statutory urban neighbourhood councils in England, as well as the continuation of rural parish councils. In addition, a number of research studies were carried out on public attitudes and the feasibility of local councils in preparation for local government reorganisation. An association was also formed with the sole intention of pressing for the mandatory establishment of urban neighbourhood councils. As spatial responses to processes of resource allocation in local areas, such observations can reflect three major dimensions: the resource needs of any given area; the social characteristics of the local population; and the area's relationship with the local authority.

Since 1961, the Association of Municipal Corporations and the then National Association of Parish Councils had advocated the establishment of urban counterparts to rural parish councils. The statutory absence of such bodies was seen to be anachronistic, since rural districts, which had become 'urbanised' by the spread of suburban housing, had retained their parish councils by virtue of having rural status. Two major arguments, however, could be voiced against such innovations. First, while local councils with minimal powers might be suitable for sparsely populated rural areas, more densely populated 'urban parishes' would be far more dynamic and hetero-geneous, with wide-ranging and often conflicting demands. Differently con-stituted areas would be required to meet different needs. It would thus be difficult to establish static boundaries. Implicit in this argument is the fact that individual demands, desires and values change over time according to age, occupation, income, family status and degree of social integration, and that such changes at the local group level can be better articulated by the activity and functions of local voluntary organisations.

Secondly, urban local authorities have argued that adequate voluntary associations already exist and co-operate with urban authorities in such a way as to render any local formal body superfluous. Yet it is often the case that local authorities take over and operate successful voluntary services themselves. Voluntary bodies tend to lack political weight, and the local authority is usually in an ideal position to choose which particular bodies to consult and to take from them just those parts that suit it. Whenever the voluntary body makes inconvenient demands, the authority is in the position of being able to say that the association has no authority to speak for local people. Until formally elected therefore, voluntary bodies have no means of demonstrating their representative claim over any specific area. Moreover, lacking authority and with insufficient funds, voluntary bodies very easily become client organisations of the local authority, being dependent upon them to get things done. Under these circumstances, reluctance to express any justified criticism of the authority may result:[14]

> often neighbourhood organizations are neutralised by absorbing them or their leaders into the bureaucracy. Alternatively, groups without upper-middle class status, values and habits of speech and argument are ignored or pushed aside in negotiation.[15]

At present, there exists a far from comprehensive and uniform pattern of urban neighbourhood councils in England. This is particularly unfortunate for two basic reasons. In the first place, these bodies are experiencing many of the shortcomings of other approaches. Like many voluntary associations, some are dependent upon the goodwill and proclivities of the local authority for finance, information and areas of potential activity. As with recent administrative approaches, neighbourhood councils, by their very name are area-based and discriminatory, with their establishment often resulting as much from external as internal forces. Secondly, in the short term, such finance, information, and other benefits which local authorities and other bodies may bestow on non-statutory urban neighbourhood councils may simply be buttressing inequalities by reinforcing the 'inverse care law' (see p. 180). Within this perspective, existing neighbourhood councils in England may do little to ameliorate the imbalance in the articulation of local area aspirations within our urban society. On the other hand, if they were to be comprehensively established, they might well help mediate local conflicts, articulate local consensus where it exists, and reflect the differing resource needs within the urban area as a whole.

Major dimensions of the roles of local councils, actual or potential, can be discerned as including: community development and community action; acting to foster a sense of community responsibility and helping those in need of special facilities (ranging from helping to provide children's play space, and alarm systems for the elderly, to the management of local festivals and celebrations); and community representation. The last-mentioned involves a four-fold process: ascertaining the views and the needs of the local area and identifying local problems and issues; monitoring those activities of governmental, public and private bodies likely to affect the local area adversely or otherwise; co-ordinating local views and aspirations, and deciding upon priorities which often involve potential conflict between individual or group interests; and expressing these views and aspirations to various levels of government and public bodies.

Whichever role any council undertakes would depend upon the area's needs. These in turn reflect such local characteristics as social and morphological composition, the nature and extent of existing local organisations, and perceived and actual access to limited resources as expressed in the area's relationship with the local authority. All of them evolve and develop over time and all are potential components of the 'valued' environment. However, to encourage the establishment of local councils for urban areas as an implicitly 'good thing' without considering local conditions and the principles outlined above, could actually decrease the likelihood of public participation. In the first place, neighbourhood councils can become dominated by politically organised and articulate residents at the expense of those without the experience or self-confidence to organise themselves. One, therefore, can see the importance of arriving at socially homogeneous areas for such councils. Secondly, they may emphasise too heavily one particular aspect of local affairs and attract support from people solely interested in that at the expense of others. Thirdly, statutory neighbourhood councils could become microcosms of local authorities themselves, compartmentalising individuals' problems, and farming out responsibility for them to different committees. Fourthly, such bodies may divert already inadequate resources away from community development while merely becoming 'talking shops' rather than acting as bases for stimulating local activity. Fifthly, although neighbourhood councillors are theoretically more in touch with their constituents, it can be argued that committees of all types seek to increase their status and influence. A neighbourhood council could find itself representing only the views of neighbourhood councillors themselves. Lastly, the existence of 'representative' neighbourhood councils could inhibit and divert local authority attention away from alternative, more flexible or experimental forms of community representation.

In discussing the environmental dimensions of locally based councils, the need to articulate a local or community identity and to relate this to a local socio-economic functional area recurs in the literature. Several surveys have been carried out at varying scales asking residents to draw or express what they consider to be their local or home area.[16] Yet, irrespective of the problems involved in the recognition, representation and translation of these cognitions, their relevance to the establishment of local councils remains arguable. In terms of relating these cognitive maps to functional areas, the Department of the Environment[17] suggested that a local council's area should have four basic characteristics: a population of 3000 to 10000; a clear spatial definition; to be regularly regarded as a particular and separate district; and no controversy over its boundaries.

We thus return, inevitably, to the question of attempting to define – both semantically and spatially – 'valued environments'. Elsewhere the present writer has attempted to present a conceptual framework for the delimitation and evaluation of such areas.[18] If, however, residents are to be able to articulate the needs and aspirations of local areas – values which may change dramatically both in time and space – any representative framework for participation within the planning process will need to be implicitly flexible and amenable to the proclivities of local conditions. Although flexible, voluntary associations may not be able to express, or be seen to express areal representation while formalised, spatially based neighbourhood councils may lack structural flexibility. It is perhaps still too early to learn from Scottish experience with community councils,[19] but the residents of local 'valued environments', by the very ambiguity and vagueness of that term, may find such a concept too elusive for the realities of the politics of planning.

Notes

1 Burnett, A. D. 1975. *Areas for Statutory Neighbourhood and Community Councils in Cities*, 6–7. Department of Geography, Portsmouth Polytechnic.
2 Spencer, D. 1973. *An Evaluation of Cognitive Mapping in Neighbourhood Perception.* Research Memorandum 23, Centre for Urban and Regional Studies, University of Birmingham.
3 Geographia, n.d. *Portsmouth, Southsea and Cosham: Large Scale Detailed Street Plan.* London: Geographia Limited.
4 Everitt, J. and M. Cadwallader 1972. *The Home Area Concept in Urban Analysis.* Paper presented at the Environmental Design Research Association Conference.
5 For example Lee, T. R. 1968. Urban neighbourhood as a socio-spatial schema. *Human Relations* 21, 241–68.
6 Styles, B. J. 1971. Public participation – a reconsideration. *J. Town Planning Inst.* 57, 163–7.
7 For example Broady, M. 1968. *Planning for People.* London: Bedford Square Press.
8 Janowitz, M. 1951. *The Community Press in an Urban Setting.* Glencoe: Free Press.
9 Baril, L. 1971. L'image urbaine. *Recherches Sociographiques* 12, 227–37.
10 Bernstein, B. 1958. Some sociological determinants of perception. *Br. J. Sociol.* 9, 153–174; *idem.* 1959. A public language: some sociological implications of a linguistic form. *Br. J. Sociol.* 10, 311–26.
11 Goodchild, B. 1974. Class differences in environmental perception: an exploratory study. *Urban Studies* 11, 157–69.
12 Duncan, S. S. 1974. Cosmetic engineering or social planning? *Area* 6, 259–71.
13 Hall, D. R. 1978. *A Geographical Study of Social Divisions in Portsmouth.* Unpublished Ph.D. thesis, University of London.
14 Baker, J. and M. Young 1971. *The Hornsey Plan,* 3rd edn. London: Association for Neighbourhood Councils.
15 Simmie, J. M. 1974. *Citizens in Conflict,* 221. London: Hutchinson.
16 Research Services Limited 1969a. *Community attitudes survey: England.* London: HMSO.
idem. 1969b. *Community attitudes survey: Scotland.* London: HMSO.

Hampton, W. 1970. *Democracy and community: a study of politics in Sheffield*. London: Oxford University Press.

Bryant, D. and D. Hall 1971. *Neighbourhood Councils in Brentwood: a Feasibility Study*. Department of Geography, Polytechnic of North London.

17 Department of the Environment 1974. *Neighbourhood Councils in England*. Consultation Paper. London: Department of the Environment.

18 Hall, D. R. 1977. Applied social area analysis. *Geoforum* **8,** 277–311.

19 Burnett, A. D. 1976. Legislating for neighbourhood councils in England. *Local Government Studies* **2,** 21–38.

11 *Environmental pluralism and participation: a co-orientational perspective*

DAVID L. UZZELL

The conflicting perceptions and interpretations of both urban and rural environments have been a major subject for attention throughout the development of environmental psychology.[1] The conclusion that can be drawn from many of the urban studies is that the city comprises a mosaic of perceptual worlds which touch, but rarely interpenetrate, each other. In this essay it is suggested that individuals and groups operate not in isolation but in relation to others; people with whom they may not physically interact but who are nevertheless significant in their lives both in terms of influencing their conceptions of the world and the part they can play in changing or conserving the built environment. Consequently a basic question is posed: do urban residents believe there is any congruency between their world and the way they believe planners and politicians construe their world? Likewise, what is the degree of similarity between the way decision makers perceive and interpret the environment and the way they think residents view the same environment?

The subject of this chapter is the application of a co-orientation model to an analysis of the plurality of perspectives on an inner-urban environment held by different urban groups. It seeks to examine the degree to which these interpretations of reality are reciprocated and to investigate the relationship between the different groups' interpretations of reality and social and political action. The essay begins with a detailed discussion of the co-orientation technique in the context of the essentially social model of man which it assumes: a model notably absent from many studies in environmental psychology. After briefly describing the study that provided the data for this chapter, there follows an analysis of the communication channels between those in government and the governed, and the level of political knowledge of the electorate. This not only raises interesting political questions in itself but also provides a persuasive argument for adopting a co-orientational approach to the analysis of political and environmental communication. The chapter concludes by examining the application of a co-orientation model to one specific environmental problem for the residents of an inner-urban area.

Co-orientation and a social construction of reality

Comparing the perceptions of different urban groups is only the first step towards explaining how those groups interpret the environment and how their interpretations influence the political process. It has to be recognised that environmental perceptions are the product of an interactive process between groups. For example, local councillors' perceptions of an area are not formed in isolation; they are the product of an interaction between their own perceptions and the perceptions they believe others to hold. In short, to understand how urban groups perceive the environment the analysis must move away from individual interpretations and comparisons to an appreciation of the *social* nature of perception process.

There is a tendency to think of the environment only in physical terms: a solipsistic reality of places and spaces which require only sensory or technical apparatus and a single individual to determine its existence, identity and dimensions.[2] However, for much of the environment to become meaningful, our understanding of reality demands not so much physical confirmation as social interpretation. Meaning in the environment is the product of beliefs, values and attitudes, and is not amenable to the same type of validation. In such cases we turn to others to assist our judgements. Reality becomes social as it is a product of group communication and negotiation; it is accepted as those around us are seen to share the same information and ideas as ourselves. To use Moscovici's example, to ascertain the degree of democracy in a country presupposes a collective consultation and agreement among the members of the group.[3]

A second perspective on this approach is provided by Berger and Luckmann,[4] who suggest that reality is a construction of society, that the subjective interpretation of the world held by others, especially 'significant others'[5] becomes objectively available, in a phenomenological sense, to an individual. This does not necessarily mean that one person understands another, or that there is any degree of congruency between one and the other's subjective processes; their subjective interpretation of the world is, however, made apparent. The individual in turn assimilates that subjective reality or at least his interpretation of the world around him, which may or may not be modified in the process. For Berger and Luckmann, the result is that 'in the complex form of internalisation, I not only "understand" the other's momentary subjective processes, but I "understand" the world in which he lives, and that world becomes my own.'[6] As the authors conclude, 'internalisation in this general sense is the basis, first, for an understanding of one's fellow men and, secondly, for the apprehension of the world as a meaningful and social reality.'[7]

This discussion is relevant both for the theoretical propositions and methodological implementation of the co-orientation model, and for the extension of participatory practices and ideals. For example, when co-orientation is applied to urban planning, an attempt is made to ascertain what perception politicians have of the inner city resident's world, and whether they can put themselves 'into the shoes' of the resident and thereby make the resident's world become their own. It has been argued that full membership of society can only be attained when each participates in the other's being and attains this degree of internalisation.[8] The notion that the individual is not a full member of society until he participates fully in its dialectic is an important idea as far as developing a theory of participation is concerned. If one is to advocate the creation and development of a participatory democracy in which as many citizens as possible are actually involved in the decision making process, then one should be encouraging communication between planners, politicians and the public in order that they may learn of, and from, each other. In this way, individuals and groups not only reciprocally assume each other's roles and perspectives but also each other's social world, thereby becoming full members of society.

The co-orientation model

To ascertain the social reality of an urban group, three kinds of information are required. In the first place there has to be an assessment of an individual's cognitive map of the environment, i.e. knowledge of how the individuals of a particular group construe any given situation. Next, there is a need to determine who else is relevant and significant to the group members in their conception of a situation. When these significant individuals or groups have been identified, their own conception of the situation has to be elicited. Finally, in order to demonstrate the essentially social nature of the first two propositions, the degree of correspondence between the two constructions of reality has to be assessed.

The co-orientation technique provides a useful framework for tapping into the social reality of different urban groups. The three concepts central to co-orientation are similarity (Fig. 11.1a), congruency (Fig. 11.1b) and accuracy (Fig. 11.1c). *Similarity* is the relationship between the cognitions held by two people concerning an object. In Figure 11.1, for example, both the urban resident and the elected member focus their attention on the environmental problems of an inner-urban area, Friary Ward. The first co-orientation concept represents the degree of similarity between the respective cognitions of the resident and the elected member.

The second concept, that of *congruency*, is a measure of the degree of similarity between one person's cognition of an object (e.g. the resident's cognition of the environmental problems of Friary Ward) and his perception of another's cognition of that object (e.g. the resident's perception of the elected member's cognition of the environmental problems). A high degree of congruency would suggest that the resident believes that the elected

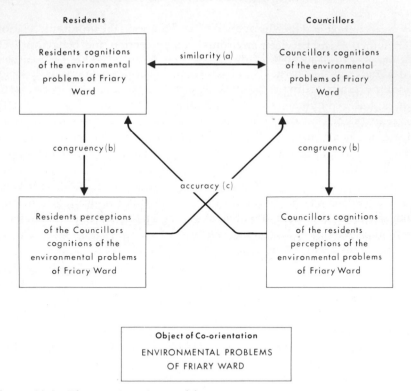

Figure 11.1   The co–orientation model.

member sees the environmental problems of Friary Ward in a similar way to himself.

The third concept is that of *accuracy*. This is the degree to which one person's estimate of another person's cognitions matches what that other person really does think. In the context of Figure 11.1, it is a measure of how accurately a resident perceives the perspective of the elected member.

It should of course be recognised that co–orientational relationships are reciprocal: the viewpoint of the elected member could equally have been taken as the starting place, and in most co–orientational studies both perspectives would be examined. Furthermore, while it is possible to study the cognitions and perceptions of each co–orientating group for the information each provides, for a true co–orientational analysis the focus of attention shifts from cognitions and perceptions *per se* to examining the relationships *between* the cognitions and perceptions.

A number of criticisms can be levelled at the individualistic approach which characterises so many environmental perception studies. Such an approach treats perception and attitude formation as a product of a person's private cognitive construction of the world;[9] it is reductionist, as the individual is taken as the unit of analysis regardless of the social, economic and political context in which he operates;[10] and it ignores the important relationship which exists between environmental perception and behaviour. The existence of similar or contrasting perceptions between two individuals does not necessarily imply behavioural consequences. The importance of the co-orientational approach lies in its recognition that a person's behaviour is influenced to a greater or lesser extent by 'his perception of the orientation held by others around him, and his orientation to them'.[11] Moreover, other factors may be brought to bear: 'under certain conditions of interaction, the actual cognitions and perceptions of others will also affect his behaviour'.[12] The relational approach central to co-orientation goes some way towards explaining not only the perceptions and attitudes of individuals but also subsequent behaviour. The behavioural consequences of co-orientational relationships will be explored in the case study discussed later.

A consensualist approach to urban and social change suggests that problems are simply malfunctions which can be cured by adjustments and rearrangements within the prevailing system.[13] The strategy employed to cure such problems would be management- or administration-oriented. One such strategy might be to improve communications between the public and the planners, as disagreement is viewed simply as the product of misunderstanding.

An alternative perspective informs us that there are fundamental differences in the way the public and the decision makers see the environment and define not the solutions, but the problems. In this instance, a conflict model of urban and social change is more appropriate. Such a model does not assume that environmental problems are perceived equally by all, nor that there is one social reality. Instead, problems are the products of value and interest differences between various sections of society.

A discussion of the co-orientation model in the context of political communication could imply that the approach is consensualist. By proposing that different environmental perspectives are the result of communication failure between the electorate and its representative(s), it might be suggested that communication difficulties need only be identified and improved for urban conflicts to be resolved. Although co-orientation can reveal conflicting perceptions between urban groups, it does not assume that the resolution of such conflicts will necessarily stem from improvements in communication, as will be seen later.

Friary Ward: an introduction

Studies of public participation in planning and the conflicts between planners and planned have invariably focused on the problems of the inner areas of large cities.[14] Guildford, a quiet and comfortable cathedral city nestling in the heart of stockbroker Surrey, must seem far removed from the grime and

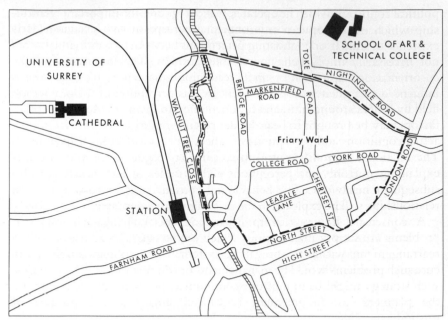

Figure 11.2   Friary Ward and the town centre of Guildford.

bulldozers of the metropolitan city, and seemingly an unlikely place for a case study of urban change and planning conflict.

Although the magnitude and severity of the environmental problems facing Guildford are obviously not comparable to those found in the larger British cities, considerable social and environmental changes have taken place over the last decade, especially in the inner-city-area known as Friary Ward (Fig. 11.2). The incursion of offices into residential areas, property speculation, pollution, blighting, housing dereliction and the sacrifice of residential property to road improvements are all features of the Guildford townscape. Furthermore, the role the public has to play in directing (and controlling) change and making it congruent with popular aspirations, needs and wishes is no less an issue in Guildford than it is in inner London. The residents of Friary Ward feel as angry as their counterparts in Rye Hill[15] with regard to the way the area has, in their eyes, deteriorated. Streets have been knocked down and familiar sights which constitute home and place destroyed in the pursuit of an ideology of economic growth.

Friary Ward is an area of primarily late 19th- and early 20th-century housing. Much of this housing is substantial redbrick, semi-detached accommodation, with a smaller quantity of terraced housing. A high proportion of these properties have been modernised with the aid of home improvement grants. The existence of a General Improvement Area (Stoke Fields GIA) containing 40 per cent of the occupied dwellings in the ward has both encouraged grant applications and made the area an attractive proposition for young first-time buyers. The area, however, has not been subject to gentrification. The local population is largely composed of two groups: elderly people who have lived in the area, if not their present houses, for a substantial part of their lives; and young couples, either childless or with a young family. Nearly three-fifths (57.6 per cent) of the residents fall within the 16–34 and over-60 age groups (compared with the figures for Guildford Municipal Borough – 47.5 per cent, and England and Wales as a whole – 44.6 per cent).[16] The majority of the households live in privately rented accommodation (53.6 per cent), although owner-occupiers are still well represented (41.1 per cent), and only a small amount of accommodation is provided by the local authority (4.9 per cent). Like most inner urban wards, the existence of large three- or even four-storey Victorian houses, easily subdivided into flats, provides homes for the more transient sections of the town's population. Most of the rented property, however, is in the form of houses which have been lived in for many years by the same family. According to the 1971 Census, 54.4 per cent of the population falls into Social Class III (manual and non-manual), which is comparable to the average for Guildford as a whole (51.7 per cent) and for England and Wales (53.5 per cent), but only 7.7 per cent of the ward population fall into Social Classes I and II (compared with 29 per cent for Guildford and 18.4 per cent for England and Wales).

Two groups of residents in Friary Ward were interviewed during the winter of 1975–6:[17] 62 members of the Friary Ward Residents Association and 75 non-members. In addition, 34 Guildford Borough Councillors (81 per cent of the full council) were interviewed, as were 14 senior Borough Council officers and other council employees involved directly with housing and planning provision in the ward.

Environmental and political knowledge

Under a representative system of democracy, politicians are elected to represent the views and interests of residents of particular areas. Nevertheless, once elected, the representatives acquire a collective responsibility for making decisions that not only affect their own ward but also other areas of the town, areas about which they may have scant knowledge. The experiential knowledge of a councillor's own ward and electorate may be considerably limited if he does not live or work in that ward. In the Guildford study it was found that councillors representing ostensibly working-class wards are much less likely to live in the wards that elected them than if they represent a higher status, more affluent ward.[18] Even Labour and Liberal councillors are more likely to live in high status wards, ironically usually represented by Conservative councillors.

When personal experience of inner city living is limited, decision makers have to rely on secondary sources of information. There are a number of means by which councillors may learn of their constituents' concerns: personal contact, informal meetings, communications by letter and telephone, local residents' group letters to the press, information via local political party machinery and sources internal to the local authority itself.[19] Furthermore, studies have shown that decision makers are selective in their sources of information.[20] The members of the electorate who utilise such means represent only a limited section of the population. Such methods of communication favour the articulate, self-confident, 'joining' sections of society. As Miller and Stokes emphasise: 'even the contacts he [the politician] apparently makes at random are likely to be with people who grossly over-represent the degree of information and interest in the constituency as a whole.'[21] Thus the reality reflected is a partial one. Furthermore, councillors rely heavily on the public coming to them; their role is passive and responsive rather than active and zetetic.

One might imagine that it would be easier for the public to know the policy perspectives and preferences of decision makers. The pronouncements of politicians in the press, in the Council Chamber and at public meetings, along with personal contact are all part of everyday political communication. Furthermore, it may be considered that the external realities of political decision making in the form of the built environment provide tangible evidence of politicians' and planners' construal of the world. In practice, of course, relatively few members of the public have personal contact with either councillors or council officials, while as yet we know little about how sophisticated an analysis people can make of the environment and political ideologies simply from the built form. Although a resident may speak to a local councillor when he carries out his triennial door-to-door electoral campaign, usually the only other opportunity of expressing views or concerns to an elected representative is at a public meeting before or after a council meeting, or by making an individual appointment. In the case of local authority officials such as planning officers, appointments are again usually the only means of access, and even then often with middle- or low-ranking officers or clerks.

In the Guildford study, residents were asked whom they had contacted in the local authority on the last occasion that they had a problem or wished to express an opinion. Only a small minority (10.9 per cent) had contacted their elected representative, while four times as many (44.5 per cent) had contacted a council officer. When residents were asked about the last time they had contacted someone from the local authority, over one-quarter (27.7 per cent) replied that they had never been in touch with anyone from the council, while another quarter (25.5 per cent) had last communicated with the council over six months prior to the study.

Another perspective on this issue is provided by the knowledge that the electorate have of their own councillor and the party he represents. Less than one-third of the residents (30.7 per cent) managed to name correctly at least one out of the three councillors who represent the ward. Over half the residents (55.5 per cent) said that they had no idea at all of the name of their councillor. Residents were slightly better informed when asked to name the political party represented by their councillor. Two in every five residents (40.9 per cent) correctly named the political party of at least one of their representatives.

Communication between the electorate and the council is at best sporadic, with no machinery in existence for the constant monitoring of public opinion. Although councillors and officers have access to a number of communication networks, especially those involving political parties and community groups, when it comes to learning about the attitudes of the majority of individuals, decision makers rely very much on the initiative being taken by members of the public. One might imagine that the arbitrary, infrequent, and highly selective collection and communication of information and opinions by councillors, officers and the public is the norm. Furthermore, given the poor state of knowledge held by the public as to whom their representatives are, such findings are hardly surprising.

Despite such erratic contact and communication, councillors and planning officers must nevertheless take decisions on the future planning of the environment. In the context of both the findings and arguments above, it would seem reasonable to assume that given an absence of information and personal experience, decision makers must somehow guess the problems and preferences of residents. At the same time there is every evidence to suggest that the public also lack information with which to make rational and coherent electoral decisions. Probably very few members of the public can articulate the policy preferences of their councillor on local issues. The widespread suspicion that local elections are significantly influenced by national party political considerations has empirical support.[22] Consequently, the elector is placed in a similar position to the elected: he must somehow guess the political and environmental orientations of the candidate.

A case study: a residents' parking scheme for Friary Ward

In 1970, Guildford Borough Council estimated that within the Stoke Fields General Improvement Area approximately 15 per cent of the residents owned at least one car. There is little reason to suspect that car ownership percentages for the whole Ward would be substantially different from the GIA figure, yet according to the 1971 Census, some 36.7 per cent of the households in Friary Ward owned at least one car. Therefore the council's figures, after just one year, were wildly inaccurate. The council, furthermore, predicted that, by 1980, car ownership in the GIA would rise to only 25 per cent of the population, arguing that 'as the average incomes in the area are at present probably lower than the national average and will remain so, car ownership rates will correspondingly not increase as quickly as the national rates.'[23] In my sample survey completed in 1976, a car ownership figure of 64.2 per cent was found, again greatly in excess of the council's projected figure for 1980. The socio-economic profile of this sample bears favourable comparison with the parent population, and similar high percentage increases in car ownership over this period have been noted elsewhere.[24]

The high car ownership levels, along with the absence of adequate off-street parking facilities, has placed a considerable strain on the amount of on-street parking currently available. In some cases this is further limited by the narrowness of the roads which permit parking on one side only; one road is only 4.65 metres wide. The problem is also compounded by Friary Ward's proximity to the two railway stations in the town and to the highly developed shopping and business centre. Consequently, a substantial number of commuters not only leave their cars in the ward during the day-time while at work in London, but the town centre shop and office workers also park there, along with shoppers and other consumers of the town's services. This particular parking problem was recognised by Guildford's planning officers back in 1970. In their proposals for environmental improvements in the GIA they suggested that: 'the arranging of landscaping and parking bays properly laid out will regularise and cut down on the amount of parking in the area where commuters, in many places, park along both sides of the road all day. Parking bays are intended primarily for the use of the area's residents and their visitors.'[25] Although the council appreciated that there was a parking problem, no proposal was put forward which would give parking priority to residents: the council simply hoped that only residents would use the available parking spaces.

As an illustration of the parking problem, every road in Friary Ward has restricted parking (single and double yellow lines) interspersed with parking bays. These bays can be divided into two types: those in which parking is limited to two hours, and those in which unrestricted parking is allowed. Unfortunately, a measure such as this is non-discriminatory; residents are penalised as much as non-residents.

In the interviews with the members and non-members of the Friary Ward Residents' Association, residents were asked what they considered to be the chief environmental problems of the area. Guildford Borough councillors were also asked to perform the same task: to articulate their cognitions of the environmental problems of Friary Ward, and, in addition, to give their perceptions of the cognitions held by residents.

Over a third (36.5 per cent) of the residents considered the parking problem to be the major environmental problem of Friary Ward. The non-members of the Ward Residents' Association considered it to be the most important problem in the area, while for the members of the Friary Ward Residents' Association (hereafter FWRA) it was second only to the dangers and pollution caused by through-traffic passing along the ward's residential roads (Table 11.1).

Table 11.1 Residents' parking problem: cognitions and perceptions of residents and councillors.

	Cognitions (1) (per cent)	Perceptions (2) (per cent)	Sample size
FWRA members (a)	31	5	62
non-members (b)	41	15	75
all residents (c)	36.5	10.2	137
all councillors (d)	38	29	34
urban councillors (e)	60	47	15
rural councillors (f)	21	16	19

It was argued earlier in the chapter that the simple comparison of awareness levels does not take us very far in terms of either predicting or explaining subsequent behaviour. The residents' parking problem provides a good example. As Table 11.1 indicates, the percentage of residents who believed that the council held a similar view to their own fell dramatically over this issue. In the case of FWRA members, only 5 per cent believed that councillors would say parking was a problem for residents. The low congruency score is immediately apparent in Table 11.2, for each of the residents' groups.

In the context of these findings, it becomes easier to explain the actions of FWRA over the residents' parking problem. As FWRA believed that few councillors were aware of the problems facing residents, the obvious course of action was to explain their problem to those who have the power to deal with it. Thus, initially, FWRA saw the issue as one of communication: if only the council could be made aware of the problem, then it would be but a short step to initiating a discriminatory policy in favour of the residents. To this end, FWRA undertook a survey to ascertain how many residents were severely affected by the parking restrictions (i.e. those residents living in roads with a two-hour parking limit). Furthermore, the Chairman of FWRA made a number of inquiries about residents' parking schemes currently in operation in Surrey and London, in an attempt to find one that might be applicable to Friary Ward. The Committee of FWRA gathered together the facts, drew up a scheme which they considered suitable and acceptable to the council, and submitted it.

Table 11.2 Co-orientational relationships: the residents' parking problem.

	Residents with councillors			Councillors with residents		
	FWRA members	Non-members	All residents	All councillors	Urban councillors	Rural councillors
similarity	−7%	+3%	−1.5%	+1.5%	+23.5%	−15.5%
	(1a–1d)	(1b–1d)	(1c–1d)	(1d–1c)	(1e–1c)	(1f–1c)
congruency	−26%	−26%	−26.3%	−9%	−13%	−5%
	(1a–2a)	(1b–2b)	(1c–2c)	(1d–2d)	(1e–2e)	(1f–2f)
accuracy	−33%	−23%	−27.8%	−7.5%	−20.5%	+10.5%
	(2a–1d)	(2b–1d)	(2c–1d)	(2d–1c)	(2e–1c)	(2f–1c)

The Residents' Association suggested that parking permits be issued to residents, and that holders of these permits should be allowed to park all day in a two-hour parking bay. This would avoid the constant requirement of finding a new parking space every two hours. The residents were still prepared to compete for parking spaces but considered that once they had found a space in their road they should not be required to move.

The co-orientation findings in Tables 11.1 and 11.2 reveal that for the council the residents' parking issue was not a problem of communication. As a collective body a substantial proportion of the councillors (38 per cent) were aware of the difficulties faced by residents. Urban councillors were particularly cognisant of the problem. Moreover, many councillors (20 per cent), especially those representing urban wards, thought that residents would consider parking to be a major problem. The Guildford Borough councillors as a whole thought that residents would consider it to be the third most important problem they faced. The urban councillors thought that the residents would give it top priority.

The figures show that residents were very inaccurate in predicting the response of the councillors to this issue. FWRA members had even less faith in the council being aware of their problem than did non-members of FWRA. The councillors, on the other hand, and especially the urban councillors, were much more accurate in their perception of the residents' concerns.

Three important conclusions can be drawn from these findings. First, FWRA's communication campaign should have been useful in informing the rural councillors of just one of the problems of inner-city living. Secondly, it would also have served the function of informing the residents themselves that FWRA were actively trying to resolve the problem. Third, for the council as a whole, the problem was not simply one of communication. The council was fully aware of the issue, as its response to the residents' proposals makes clear.

The council made it apparent to residents that if a parking scheme were to be introduced it would have to be 'economical', i.e. it would have to pay for itself and this would involve charging a cost-effective rate for each permit. Further discussions were held with the council to ascertain the cost. The Chief Constable of Surrey County Constabulary insisted that if a residents' parking scheme was to be introduced it would need at least two additional traffic wardens to police it and this added substantially to the cost of the scheme. This demand was inexplicable as the scheme applied only to those areas which were currently being policed by traffic wardens, and therefore no extra work should have been entailed. The council stipulated that the residents would have to pay all the costs of introducing the scheme (e.g. advertising, changes in traffic regulations, signposting and administrative costs), and furthermore, these costs would have to be paid in the first year – they could not be spread over, say, a five-year period.

Despite these considerable constraints, FWRA carried out a further survey to ascertain how many residents would be interested in the proposed scheme. Although the cost of each parking permit would be dependent upon the total number of participants, FWRA had to give the residents some idea as to the probable cost, with the conservative estimate that each permit would cost about £15 per annum. The response to the survey was poor, with many residents complaining that the cost was too high. With the response FWRA received, the *per capita* cost would have been considerably higher. The scheme was abandoned.

Guildford Borough Council has in recent years built three multi-storey car parks, a substantial proportion of the finance for which has had to be borrowed. There are, in addition, a number of surface car parks. All these car parks form an important element in the infrastructural support to the commercial and business sector that the council has been encouraging and promoting in the town. In order to pay the high loan charges on the money borrowed to build the car parks, Guildford Borough Council has been forced into imposing very high car parking charges, much higher than in the surrounding towns. These charges have received a bad 'press'. If a residents' parking scheme was to be introduced in inner-urban residential areas it would reduce to a negligible amount the free parking available to those people who live in the suburbs, the rural areas and beyond – in fact, those very areas where the controlling Conservative party receives the overwhelming majority of its electoral support. A residents' parking scheme would force shoppers and officer workers to use the expensive car parks; thus, it is argued, damaging Guildford's image and attractiveness as a commercial and shopping centre. Such an action would directly challenge the ideology of economic growth and expansion which Guildford Borough Council actively promotes.

Conclusion

It should be apparent that the simple comparison of environmental percep-tions by different urban groups would not have provided a great deal of insight into the participatory behaviour of FWRA. The efficacy of the co-orientation approach is borne out by FWRA's actions, which, to paraphrase McLeod and Chaffee, were influenced by its members' perception of the orientations held by others around them, and their orientation to them.[26] Co-orientation is useful for not only accounting for residents' different interpretations of the world, but also their actions as a result of those interpre-tations. Thus a closer link is forged between environmental perceptions and behaviour. Given the perceptions of FWRA, it was an altogether rational strategy to convey its members' concerns to the council, especially consider-ing the amount of contact and communication which exists between the local authority and the electorate. Furthermore, the criticism of reductionism which is sometimes levelled at the behavioural approach has less validity in the case of co-orientation, as environmental perceptions are grounded more firmly in social and political intergroup relations and action.

It was argued earlier that co-orientation need not be seen as being consensualist. It is quite apparent in this case study that communication alone between the urban groups was insufficient in resolving the conflicts. FWRA's action can only be understood in terms of its co-orientational relations with the local authority. The residents' proposals were not implemented because they failed to attract political and economic support – the amount of information was adequate. Guildford Borough Council not only objected to the residents' parking scheme on economic grounds, but also because it challenged, in a small way, their ideological orientations. The objections of the police were of a different order. Although the Chief Constable used economic and manpower arguments, it was known to both councillors and residents that he was opposed to residents' parking schemes on principle. Against such an opposition, it was doubtful whether FWRA could succeed. In this exercise in participation, the council set the ground rules with which FWRA was to operate. By not getting the support necessary to pay for a residents' parking scheme, failure was placed firmly at the door of the residents. With the council (and the Surrey County Constabulary) setting the rules, it could never be participation.

It is also apparent from this study that while ideological differences are of crucial importance in determining the success or failure of participatory exercises, communication is still nevertheless an important aspect of participation. This study demonstrates the poor level of knowledge held by rural councillors concerning the problems of inner-city living; the inadequate degree of political knowledge by residents of their councillor and the political party he represents; the poor state of knowledge possessed by planning officers of the level of car ownership in the ward; and the poor communication channels which exist between decision makers and the public. While the city as a political system has all the complexity of a silicon chip, as an information system we are still at the level of beating drums.

Notes

1 Appleyard, D. 1976. *Planning a Pluralist City*. Cambridge, Mass.: MIT Press.
   Lucas, R. C. 1970. User concepts of wilderness and their implications for resource management. In *Environmental Psychology*, H. M. Proshansky, W. H. Ittelson and L. G. Rivlin (eds), 297–303. New York: Holt, Rinehart and Winston.
2 Moscovici, W. 1976. *Social Influence and Social Change*. London: Academic Press.
3 ibid., 32.
4 Berger, P. L. and T. Luckmann 1967. *The Social Construction of Reality*. London: Penguin.
5 ibid., 151.
6 ibid., 150.
7 ibid.

8 ibid.
9 McLeod, J. M. and S. H. Chaffee 1973. Interpersonal approaches to communications research. *Am. Behavioural Scientist* **16,** 469–99.
10 Herbert, D. T. and R. J. Johnston 1978. Geography and the urban environment. In *Geography and the Urban Environment*. Vol. 1, D. T. Herbert and R. J. Johnston (eds), 1–33. Chichester: Wiley.
   Sandbach, F. 1980. *Environment, Ideology and Policy*. Oxford: Basil Blackwell.
11 McLeod, J. M. and S. H. Chaffee op. cit., 470.
12 ibid.
13 Community Development Project Information and Intelligence Unit 1974. *Inter-Project Report, 1973*. National Community Development Project.
14 For example, Cockburn, C. 1977. *The Local State*. London: Pluto Press.
   Paris, C. and B. Blackaby 1979. *Not Much Improvement: Urban Renewal Policy in Birmingham*. London: Heinemann.
15 Davies, J. G. 1972. *The Evangelistic Bureaucrat*. London: Tavistock.
16 Census, 1971.
17 Uzzell, D. L. 1980. *A Coorientation Approach to Participatory Politics in an Inner City Ward*. Unpublished Ph.D. thesis, University of Surrey.
18 ibid., 175–7.
19 Ferres, P. 1977. Improving communications for local political issues. In *Local Government and the Public*, R. Darke and R. Walker (eds), 160–79. London: Leonard Hill.
20 Dearlove, J. 1973. *The Politics of Policy in Local Government*, 187–90. Cambridge: Cambridge University Press.
21 Miller, W. E. and D. E. Stokes 1963. Constituency influence on Congress. *Am. Polit. Sci. Rev.* **57,** 45–56.
22 Green, G. National, city, and ward components of local voting. *Policy and Politics* **1,** 45–54.
23 Guildford Borough Council 1970. *Housing Act, 1969: Proposed Stoke Fields General Improvement Area*. Unpublished MS.
24 Stringer, P. 1978. *Tuning in to the Public: Survey before Participation*. Interim Research Paper 14, Linked Research Project on Public Participation in Structure Planning, University of Sheffield.
25 Guildford Borough Council, op. cit., 5.
26 McLeod, J. M. and S. H. Chaffee, op. cit.

Index

access 61, 65, 67, 174
action 21, 104, 159
action areas 180
action research 165–6
Adair, D. 90
aesthetics 7, 28, 100–23, 178
Ageé, T. 24
Aldiss, B. 82
Allison, L. 3
ambiguity in streets 154–9
amenity 100, 157
ancient monuments 6, 62–3, 71, 78, 86, 89, 93
architectural determinism 177
Architectural Review 23
Arnaud, J. J. 78
Art Deco 6
artists 2, 101, 156
Ashbrook, K. 56–7, 62, 64
attachment to past 7, 74
attachment to place 2, 3, 5, 8, 43, 161–2, 168, 172–3, 177–8

Banham, R. 11
Barrell, J. 138
Barrett, W. H. 47
beauty 3, 100–1, 108, 114, 155
Bedford 6, 10–34
Belloc, H. 44
Berger, P. L. 190
Bernstein, B. 178
Betjeman, J. 10, 35
Bradbury, R. 85, 90
Brontë, A. 63
Brontë, E. 60
Brown, L. ('Capability') 125, 127
Buchanan, C. 70
Bulletin of Environmental Education 25
Burnett, A. D. 172–3
Buttimer, A. 167

Cameron, J. 10
Canter, D. V. 162
caring for place 1, 145–60
Carney, W. 30
Carson, Robin 81
Casson, H. 162
child development 1, 4, 161–3
cinéma vérité 36
Civic Trust 10, 23, 153
Clarke, A. C. 83
cognition, *see* perception
cognitive maps 174, 191
communication 164, 189, 193, 197, 199, 202
comprehensive redevelopment 2, 155

Conan Doyle, A. 59
conservation 7, 10, 23, 55, 58, 64, 70, 71, 145–60
Conservation Areas 146, 152
Cooper, C. C. 11, 162–3
Cooper-Hewitt Museum 30
co-orientation 8, 189–203
Cornish, V. 41, 66–7, 100
Countryside Commission 68
Countryside Review Committee 69
Cowper, W. 124
creativity 1, 4, 101, 115, 145
crime 48, 80
cruelty 49, 52
Cullen, G. 21, 100, 108, 148–50

Darby, H. C. 39, 45
de Beauvoir, S. 86
Defoe, D. 44, 47
design, urban 10, 32
Design Council 155
developers 153
dialectical materialism 7, 100, 112–17
documentary films 35–54
Dower, J. 66, 67
Drabble, M. 74, 94
Dugdale, W. 45, 47
du Maurier, D. 81, 82, 84

Edmondson, G. C. 83, 85
education 10, 24, 31, 167, 168
emotion 4–5, 8, 22, 23, 77, 91, 108, 111, 113, 173
English, P. 161
environmental biographies 1, 27–8
environmental change 2, 145, 151–4, 158–9, 165
environmental determinism 47
environmental legislation 55, 65–72, 164–5
ethology 107, 109, 110
evaluation 22, 104
existentialism 106, 110, 111
experience 1, 2–3, 6, 8, 10–15, 20–1, 22–30, 37–40, 56–7, 58, 74, 80–2, 84, 91, 100, 102, 108–11, 113, 117, 118, 158, 178, 184, 195

Farmer, P. J. 83
feeling, sense of 6, 21, 23, 25, 80
Fens 6, 35–54
'Fen Tiger' 39, 46–50
Finney J. 81, 84–5, 86
floods 39, 44–6
forestry 71–2
formalism 117
Forster, E. M. 74–5
Fried, M. 164
functional tradition 150

games 25–7
General Improvement Areas 165, 167, 180, 194–5, 197–8
geography 24–5, 106, 107–8
Gerardin, R. L. 127
Gerrold, D. 81
Gestalt, laws of 103, 110
Gibbon, E. 89
Gilpin, W. 125, 127
Godwin, H. 41
Goldsmith, O. 124–5, 134
Gombrich, E. H. 107
Goodchild, B. 178
Gosling, R. 2
Guildford 191–5, 197–203
Guthlac, St 44

habitat theory 104
Halprin, L. 25–7
Hanff, H. 89
Hare Street 127
Hartley, L. P. 89
Hawthorne, N. 89
Haydon, G. 37, 52
Haythornthwaite, G. 63
Heap, D. 3
heathland 61
Helphand, K. 27
hermeneutics 117
Hobhouse, A. 66
home 161–2
Hoskins, W. G. 35, 74, 77, 79
Housing Action Areas 165
Hoyland, J. 15
Hoyle, F. 83
Husserl, E. 111
Huxley, A. 3

iconography 107
idealism 117
identity 5, 21–2, 74, 78, 101, 115, 146, 158–9, 161, 190
imagination 74, 101, 113
industrialisation 141
information 20, 22
information theory 103, 104, 107, 110
interpretation 10–34, 100–1, 102, 105, 106, 114, 189
inverse care law 180, 185
isolation 39–40, 50–1

Jackson, J. B. 107, 114–15
James, H. 76

Kingsley, C. 48

landowners 2, 124–6, 127, 140
land passion 42
land reclamation 44–6, 68, 71

landscape aesthetics, *see* aesthetics
landscape gardening 7, 127
landscape parks 124–44
landscape quality 102, 104, 108–9
learning 22
Legge, C. 46
Lenz-Romeiss, F. 5, 162–3, 164
Lewis, P. 24
linguistics 111–12
Lively, P. 75, 85
local area 3–4, 7, 161–203
local politics 181–4
Lock, M. 147, 156
Lowenthal, D. 65, 113
Luckmann, T. 190
Lynch, K. 4, 21

McCarthy, M. 80
McDonald, S. 167
MacEwen, M. 57, 60, 61, 64
McGuire, G. 57, 59–60, 65
Manplan 23
Marxism 112, 178
mass media 2, 31, 35–54
Mass Observation 11
Mather, C. 24
Mayfield, R. 161
meaning 5, 8, 39, 77, 101, 103–4, 105–11, 114
Meinig, D. 4
memory 28, 87
metaphysics 117
Middleton, A. P. 90
Miller, H. 90, 94
Mitchell, D. 36, 52
mobility 163–5, 177
moorland 4, 6, 55–73, 78
Morris, J. 10
motivation 1, 2, 110, 184
movement 21–2, 61
Mumford, L. 89, 90
mystery 77

Nairn, I. 10
national heritage areas 69
national parks 7, 55, 58, 65–72
National Trust 44
navigation 20, 21, 113
neighbourhood 2, 4, 161–203
neighbourhood councils 172
new towns 165, 177
Nin, A. 77
nostalgia 76
notation 27

opium 48, 50

parish councils 184
participation 2–3, 116, 165, 167–9, 184, 187, 189–203

peasant mentality 39, 43
Pennine Way 55–6
perception 21, 92–3, 100, 102–5, 108, 110,
    166, 173–7, 178, 180, 189–90
Percival, A. 29
phenomenology 110, 111, 117, 190
Phillips, W. 90
photography 3, 23, 27, 31
Picton, T. 23, 31
picturesque landscapes 125, 127
Plater, A. 52
poets 2
political processes 2, 20, 55, 100, 116, 156,
    164–9, 191–203
political representation 165–9, 172, 179–84,
    186–7, 196–7
Poor Laws 132, 136
Pope, A. 125, 128, 134
Portsmouth 7, 173–4, 180–4
positivism 100, 104
preferences 2, 6–7, 65, 172
preservation 4, 58, 93
pressure groups 2, 56
Price, U. 125
Prince, H. C. 65, 113
psychology 110, 189

Ramblers' Association 56
Rapoport, A. 102
recreation 10, 68
Red Books 126, 134–6, 138
reference groups 163
rehabilitation 152, 165
Relph, E. C. 158
Repton, H. 7, 124–44
resource allocation 184, 186
Robinson, D. 91
romanticism 100, 111
Ruskin, J. 90

Sandford Committee 68–9
Sanoff, H. 25
Sayer, S. 56, 57, 60, 62, 64, 65, 78
Schlesinger, J. 36
Schools Council 25
science fiction 7, 79–91
Scruton, R. 157
semiology 106, 110, 111, 117
Sennett, R. 163
sensation 21, 25–8, 39, 40, 103, 127–8, 173
sense of place 35, 51–3, 161–2
service provision 178–9
Sheringham 7, 124, 131–8
Simak, C. 80, 83–4
social engineering 156
social investigation, methods of 3, 6, 110,
    173–4, 179
socialisation 5

social networks 164
solitude 63
Spencer, D. 168
Spender, S, 88
Stephenson, T. 55–6, 57, 59–60, 63, 65, 67
stereotype 10, 27, 31, 77, 146, 153, 157–8
street furniture 148–51
structuralism 107
suburbia 116–17, 155
Swain, H. 24
Swallow, N. 37
symbolism 106–7, 109–11, 116, 173

tastes 5, 100
    elite 2, 4, 6–7, 55, 65, 106, 127–8
territoriality 104, 109, 113, 174–7
Thompson, E. P. 137–8
Tilden, F. 29, 91
Tindall, G. 77
Tinker, J. 69
topophilia 162
tourism 10, 77, 92
townscape 108–9
town trail 10, 16–20
travellers 2, 79, 82
Trevelyan, G. M. 66
Tuan, Y. F. 105, 107, 114
Tucker, W. 82

Uttley, A. 85

values 2, 5–6, 7–8, 35, 43, 100, 101, 105, 118,
    124, 147, 159, 168–9, 172, 193
Venice 22
Victorian Societies 6
vision 101, 103, 108, 114–15
voluntary organisations 180, 184–5

Wagner, P. 21
Walker, R. A. 117
Ward, C. 10
Webber, M. M. 163
Wells, H. G. 87
West Hartlepool 7, 145–60
wilderness 57–64, 68
    conditions for 59–64
Williams, R. 75, 107, 112–13, 118, 158–9
Willmott, P. 164
Wolfe, T. 24
Woolf, V. 87
Worsthorne, P. 89
writers 2, 74–99
Wurman, S. 25

Young, M. 164

Zeitgeist 106, 110
Zweig, P. 90